TECHNICAL REPORT

The RAND SLAM Program

Jacob Alex Klerman, Christopher Ordowich,
Arthur M. Bullock, Scot Hickey

Prepared for the Office of the Secretary of Defense

 NATIONAL DEFENSE RESEARCH INSTITUTE

This research was sponsored by the Office of the Secretary of Defense (OSD) and conducted within the Forces and Resources Policy Center of the RAND National Defense Research Institute (NDRI) with NDRI concept development funds, and then as part of the OSD/RA sponsored project "Sustaining the RC." NDRI is a federally funded research and development center sponsored by the OSD, the Joint Staff, the Unified Combatant Commands, the Department of the Navy, the Marine Corps, the defense agencies, and the defense Intelligence Community, under Contract W74V8H-06-C-0002.

Library of Congress Cataloging-in-Publication Data

The RAND SLAM program / Jacob Alex Klerman ... [et al.].
    p. cm.
    Includes bibliographical references.
    ISBN 978-0-8330-4212-5 (pbk. : alk. paper)
    1. United States. Army—Organization—Computer programs. 2. United States. Army—Reserves—Computer programs. 3. United States. Army—Personnel management—Computer programs. 4. United States. Army—Cost control—Computer programs. 5. Risk assessment—United States—Computer programs. 6. Military planning—United States—Computer programs. 7. Military planning—United States—Decision making. I. Klerman, Jacob Alex. II. Rand Corporation.

UA25.R36 2008
355.30285'53—dc22

2008049259

The RAND Corporation is a nonprofit research organization providing objective analysis and effective solutions that address the challenges facing the public and private sectors around the world. RAND's publications do not necessarily reflect the opinions of its research clients and sponsors.

RAND® is a registered trademark.

Published 2008 by the RAND Corporation
1776 Main Street, P.O. Box 2138, Santa Monica, CA 90407-2138
1200 South Hayes Street, Arlington, VA 22202-5050
4570 Fifth Avenue, Suite 600, Pittsburgh, PA 15213-2665
RAND URL: http://www.rand.org/
To order RAND documents or to obtain additional information, contact
Distribution Services: Telephone: (310) 451-7002;
Fax: (310) 451-6915; Email: order@rand.org

Preface

This report provides a multilevel description of the RAND SLAM program (SLAM stands for Force Structure, Force Levels, and Force Assignment Model). The RAND SLAM program was developed to help analysts think through issues about the optimal size and structure of U.S. military forces. This report is primarily intended to serve as a user's guide for analysts who wish to use the RAND SLAM program. However, the report also demonstrates some of the unique features of the RAND SLAM program for less technical readers.

The main document provides an overview of the program, primarily through a sequence of examples. Users requiring detailed knowledge of the RAND SLAM program should see Chapters One and Two of this report. Chapters Three and Four validate the RAND SLAM program by showing how the program can reproduce results from a generally accepted analysis. Chapters Five through Seven provide an overview of the unique kinds of analysis that can be performed with the RAND SLAM program. The appendixes of this report provide reference material for users of the RAND SLAM program.

This research was sponsored by the Office of the Secretary of Defense (OSD) and conducted within the Forces and Resources Policy Center of the RAND National Defense Research Institute (NDRI) with NDRI concept development funds, and then as part of the OSD/Assistant Secretary of Defense for Reserve Affairs–sponsored project "Sustaining the RC." NDRI, a division of the RAND Corporation, is a federally funded research and development center sponsored by the Office of the Secretary of Defense, the Joint Staff, the Unified Combatant Commands, the Department of the Navy, the Marine Corps, the defense agencies, and the defense Intelligence Community.

For more information on RAND's Forces and Resources Policy Center, contact the Director, James Hosek. He can be reached by email at James_Hosek@rand.org; by phone at 310-393-0411, extension 7183; or by mail at the RAND Corporation, P.O. Box 2138, 1776 Main Street, Santa Monica, CA 90407-2138. More information about RAND is available at http://www.rand.org/

Contents

Figures

Tables

Summary

This report describes the RAND SLAM program. The RAND SLAM program allows an analyst to explore the trade-offs inherent in military force structure decisions. More specifically, the program allows an analyst to examine trade-offs between cost, stress, and risk when the requirement for deployed forces varies over time. The RAND SLAM program's unique features allow an analyst to study the effects of varying military requirements on force structure decisions. Optimal force structures can vary dramatically depending on the nature of the threat. For this reason, the RAND SLAM program models contingencies stochastically, acknowledging that military requirements vary unpredictably over time and allowing an analyst to study the implications.

This report contains a number of illustrative analyses focusing on the force structure of the U.S. Army. Many of the analyses that have been performed in support of force structure decisions have been very narrowly focused (Center for Army Analysis, 1999, 2001, 2002, 2003). The power of the RAND SLAM program is that it allows an analyst to perform many different types of analysis under almost any set of assumptions. The program's primary focus is allowing an analyst to determine the lowest-cost force structure for a given requirement and stress level (e.g., the "optimal" active-reserve mix). However, the program can also perform many other types of analyses, such as determining the effect of different requirements on stress levels for a given force structure. The RAND SLAM program was designed to provide as much flexibility as possible. The user determines both the unit of analysis and the time resolution for each set of simulation runs. The program is capable of utilizing any unit of analysis—individual, company, battalion, brigade, etc.—and modeling any time resolution—days, months, quarters, years, etc. The RAND SLAM program also allows the user to move beyond the typical analysis of finding the optimal active-reserve mix: The user can create new types of forces and examine their attractiveness under varying assumptions.

This report begins with a description of how to use the RAND SLAM program and then presents results that demonstrate the uniqueness of the program. The report is organized by chapter as follows:

- Chapters One and Two provide a description of the RAND SLAM program and how to use it.
- Chapters Three and Four examine a problem—supplying stabilization forces for Iraq as of late 2004—that has already been thoroughly examined using intensive spreadsheet techniques in *Stretched Thin: Army Forces for Sustained Operations* (Davis et al., 2005). To validate the model, the results from the RAND SLAM program are compared with those found in *Stretched Thin*.

xvi   The RAND SLAM Program

- Chapter Five shows how to use the RAND SLAM program to perform a cost-effectiveness analysis of active versus reserve forces. In this chapter, we find that when the demand for deployed forces is constant, the relative cost-effectiveness of active versus reserve forces is very sensitive to restrictions on the use of each force.
- Chapter Six demonstrates the unique features of the SLAM program by simulating stochastic force requirements. This chapter examines the implications of a stochastic environment for force structure decisions. In this chapter, we find that, when the demand for deployed forces is no longer constant at a level at which both active and reserve forces are needed in every period, reserve forces are relatively more attractive. This chapter also demonstrates that, when planning for two simultaneous wars, active forces are relatively more attractive. The results derived in this chapter are very sensitive to the assumptions made about force costs and usage restrictions. These results are based on a primitive cost model and are only meant to illustrate the power of the RAND SLAM program.
- Chapter Seven illustrates the problem that motivated the development of the RAND SLAM program—choosing the appropriate mix of active and reserve forces. This chapter shows that the optimal active-reserve mix is very sensitive to assumptions about force costs and usage restrictions. It also shows that the optimal force mix can vary dramatically depending on the nature of the threat.

The appendixes of this document provide a more detailed description of the RAND SLAM program and how to use it. Appendix A provides a reference for users of the RAND SLAM program. Appendixes B and C provide technical details about the design of the program. Appendix D describes the process used to determine force assignment rules. Appendix E provides a detailed list of the components of the RAND SLAM program.

Acknowledgments

This research was begun with internal funding provided by NDRI, and specifically by James Hosek, Director of the NDRI Forces and Resources Program, and Gene Gritton, Vice President, NDRI. Subsequent funding was provided by the NDRI project "Sustaining the RC."

Many people at RAND contributed to the substantive insights that motivate the development of the simulation program and the details of the specification. Contributions of particular note include the comments of Beth Asch, Carl Dahlman (then at RAND, now at Office of the Under Secretary of Defense for Personnel and Readiness), Meg Harrell, James Hosek, and Tom Lippiatt. Comments of staff at OSD, including John Winkler and Jennifer Buck, were also invaluable.

Development of the RAND SLAM program has benefited from comments from a wide spectrum of RAND programmers, research assistants, and Frederick S. Pardee RAND Graduate School students. Assistance in bringing together the RAND SLAM programming team was provided by Jan Hanley, director of RAND's Research Programming Group, and Craig Martin, a Research Programming Group manager. RAND Project AIR FORCE supplied a LINUX workstation with a General Algebraic Modeling System (GAMS) license.

John A. Ausink and Kenneth Girardini provided helpful reviews of this report. Paul Steinberg provided editorial support, and Christopher Dirks provided help with formatting the manuscript.

Acronyms

AC	active component
DoD	U.S. Department of Defense
GAMS	General Algebraic Modeling System
GWOT	global war on terror
IP	integer program
LP	linear program
MB	Megabyte
MRC	major regional contingency
RC	reserve component
SLAM	Force Structure, Force Levels, and Force Assignment Model
VBA	Visual Basic for Applications

CHAPTER ONE

Introduction

The threats faced by the U.S. military and the resources available to address these threats continue to shift. The Cold War enemy has receded. Nevertheless, two recent wars in Iraq and a prolonged stabilization mission there have emphasized the importance of maintaining large and robust ground forces. How large those forces need to be and the appropriate mix of active versus reserve forces remain open issues.

Choosing the amount and mix of land forces is a crucial defense question. With too few forces, the nation is unable to pursue its foreign policy goals—or at least, doing so puts considerable strain on the military, its soldiers, and their families. With too many forces, the nation wastes its money or underfunds other components of the defense budget.

These decisions are difficult, in part because of their importance. However, force structure decisions are also difficult because they are made in an inherently uncertain environment. Planning for a known, fixed level of required forces would be relatively easy. However, most of the requirements for troops are uncertain. Perhaps there will be no more major conflicts for the next generation; perhaps there will be one in ten years; perhaps in two years; perhaps there will be two nearly simultaneous major conflicts.

Force structure choices need to account for this uncertainty. The appropriate response to the possibility of one war is not simply a proportionately larger force than for peacetime. These issues are even more dramatic once we consider the possibility of two or more conflicts.

Optimal force decisions require trade-offs between readiness, "force stress," and cost. This is a multidimensional problem. Heuristics are useful; careful analysis requires simulation. By modeling contingencies stochastically, the RAND SLAM program allows an analyst to explore the impact of varying force requirements on force structure decisions.

1.1. The RAND SLAM Program

The RAND SLAM program is a RAND-developed software tool that allows an analyst to explore the attractiveness of varying force structures from a dynamic perspective. The RAND SLAM program is unique in its ability to simulate varying force requirements. Previous work on this topic has assumed a constant demand for forces, which allowed the analysis to be performed in a spreadsheet (see Davis et al., 2005). The ability to simulate varying force requirements allows an analyst to explore the implications of the uncertainty in requirements for force structure decisions.[1]

This report contains a number of illustrative analyses focusing on the force structure of the U.S. Army. Many of the analyses that have been performed in support of force structure decisions have been very narrowly focused (Center for Army Analysis, 1999, 2001, 2002, 2003). The power of the RAND SLAM program is that it allows an analyst to perform many different types of analysis under almost any set of assumptions. The program's primary focus is allowing an analyst to determine the lowest-cost force structure for a given force requirement and stress level. However, the program can also perform other types of analyses. For example, it can measure the amount of stress placed on a single force structure for varying requirement levels. This is similar to the analysis performed in *Stretched Thin* (Davis et al., 2005). The RAND SLAM program is structured so that the user can perform many different types of analyses. Therefore, a complete analysis requires the user to define different simulation "runs" with varying inputs. A typical analysis will create multiple simulation runs that hold force requirements and assignment rules constant but vary force structures in order to determine which force has the lowest cost under a given set of assumptions. The range of force structure decisions is limited only by the input data provided by the user; the SLAM program does nothing to limit the options that can be explored. The RAND SLAM program also allows the user to move beyond the typical analysis of finding the optimal active-reserve mix: The user can create new types of forces and examine their attractiveness under any set of assumptions. Figure 1.1 summarizes the basic high-level structure of the model.

The analyst specifies the possible contingencies (and their likelihood), the available forces (and their readiness), constraints on the availability of each force, and a set of force allocation rules. Given this input, the program generates sequences of conflicts and applies the force allocation rules. In each period, the program records various outcomes (e.g., ability to meet the military requirements, stress on the force from too-frequent or too-long deployments, and cost). These outcomes can be summarized for one period (small N) or over many periods (large N).

In a complete analysis, an analyst would begin by specifying a set of contingencies. A simulation would be run for varying force levels and force allocation rules. The analyst would then compare outcomes (e.g., cost, stress, risk) across the runs. Finally, to assess the robustness of the results, the analysis would be repeated for varying specifications of the contingencies.

The RAND SLAM program was designed to provide an analyst with as much flexibility as possible. The user determines the unit of analysis and time resolution for each set of simulation runs. The RAND SLAM program is capable of utilizing any unit of analysis—individual, company, battalion, brigade, etc.—and modeling any time resolution—days, months, quarters, years, etc. This flexibility requires that the user be consistent with all unit and time definitions within a set of simulation runs. In all of the analyses contained in this report, we assume that the unit of analysis is a Brigade Combat Team (BCT) and that time is measured in quarters.

---

[1] It is important to note that the spreadsheet methodology used in *Stretched Thin* can be extended to varying requirements for deployed forces when periods of war and peace last a long time. However, this analysis is potentially misleading given the number of short-duration conflicts in U.S history (Grenada, Panama, etc.). These types of contingencies are important to model because they reduce the availability of forces for other conflicts and, if they occur frequently enough, require a substantial commitment of forces. The RAND SLAM program allows an analyst to move beyond steady-state calculations to find a more general solution to the force-planning problem.

Figure 1.1
High-Level Structure of the RAND SLAM Program

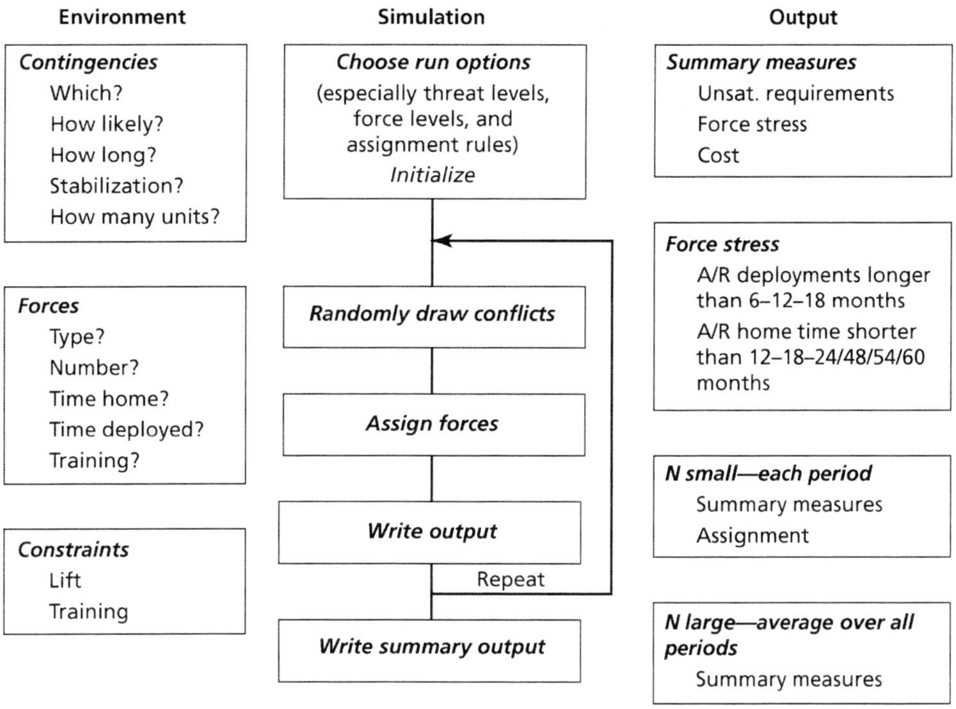

RAND TR433-1.1

As noted earlier, the RAND SLAM program provides three main outputs: cost, stress, and risk. Cost is measured on a per-unit basis. Costs differ by force and status (home, deployed, or training). The SLAM program can be easily manipulated so that costs also differ during different periods of training and deployment. Stress is measured in two ways: long deployments and short-dwell times. Long deployments are measured as the percentage of units that are deployed for longer than a specified period of time. Short-dwell times are measured as the percentage of units that are deployed with less than a specified amount of time at home. Finally, risk measures the ability of a given force to meet the simulated requirements. The RAND SLAM program can measure risk in two ways: (1) as the percentage of periods in which military requirements are not met, and (2) as the average number of units a force cannot provide over all periods. With some basic knowledge of Microsoft Access queries, the user can define other variations of cost, stress, and risk measures.

1.2. Role and Structure of This Document

This document provides a multilevel description of the RAND SLAM program. The main document provides an overview, primarily through a sequence of examples. Chapter Two shows how to open an existing SLAM database, examine the input data, execute a run, and examine the output data. Chapter Three shows how to replicate the analysis performed in *Stretched Thin* (Davis et al., 2005) using only active forces. Chapter Four shows how to create and use reserve forces while replicating further analyses contained in *Stretched Thin*. Chapter Five illustrates some unique features of the RAND SLAM program by performing a cost-effectiveness analysis. Chapter Six continues to illustrate the uniqueness of the RAND SLAM program by introducing varying force requirements. The final substantive chapter, Chapter Seven, shows how the RAND SLAM program can inform force structure decisions by examining the trade-offs of altering the active-reserve mix.

Beyond the main chapters, the document includes several appendixes. Appendix A provides a reference guide for the user. As such, it reviews the major model concepts, screens, and sequence of tasks to define and execute a simulation run. Appendixes B and C provide technical details about computer implementation and optimization algorithms. Appendix D describes the process used to determine force assignment rules. Appendix E contains a listing of the tables and modules included in the program.

CHAPTER TWO

Using the RAND SLAM Program

This chapter shows how to open and execute an existing simulation run in the RAND SLAM program. A simulation run combines contingencies, forces, and an assignment rule to create a scenario that is simulated by the RAND SLAM program. This chapter illustrates the core components of the RAND SLAM program, which provide the user the flexibility to create runs with differing threats, force sizes and structures, and assignment rules.

The first section in this chapter shows how to install and open the RAND SLAM program. The second section shows the inputs required to execute a run. The third section shows how to execute a run. The last section introduces many of the reports and graphs produced by the RAND SLAM program. This chapter is designed so that the user can follow along interactively with each step. The example used in this chapter is designed to be illustrative of the program environment; later chapters demonstrate the unique analytic features of the RAND SLAM program.

2.1. Installing and Opening the RAND SLAM Program

The RAND SLAM program distribution CD[1] contains version 1 of the RAND SLAM program. This version includes three Microsoft Access Databases, an Excel spreadsheet, two General Algebraic Modeling System (GAMS) scripts, and a Visual Basic matrix components installer. To use the RAND SLAM program, the user must first copy these files to his or her hard drive. We suggest that the user create the folder C:\Program Files\SLAM and copy all seven files to this folder before proceeding.

The three Microsoft Access databases on the distribution CD include one "front-end" and two "back-end" databases. The front-end database includes the user interface and some global program variables, but no data tables. The back-end database includes all of the data tables containing the information needed for SLAM simulation runs. The front-end database included in the distribution package of the RAND SLAM program is the file "Slam1.mdb." The first back-end database is called "SLAM1be.mdb." This file contains the very basic set of data and runs for use in this chapter. The other back-end database (SLAM1Analysis.mdb) includes all of the contingences, forces, and assignment rules used in later chapters of this report.

---

[1] To obtain the RAND SLAM distribution CD, contact RAND's Forces and Resources Policy Center using the information provided in the preface of this report.

To begin using the RAND SLAM program, the user should open the front-end database SLAM1.mdb, either by double clicking on the icon or by right-clicking on the icon and selecting *Open*. This will bring up the SLAM start-up screen shown in Figure 2.1.

Figure 2.1
SLAM Start-Up Screen

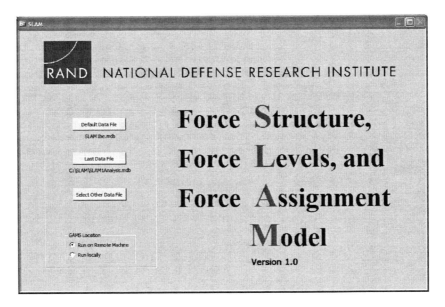

RAND *TR433-2.1*

In this chapter we use the back-end database SLAM1be.mdb. To open this back-end, click on the *Default Data File* button on the start-up screen. In the distribution version of the SLAM program, the default data file will always be SLAM1be.mdb.

After clicking on the *Default Data File* button, the user should see the SLAM toolbar, which includes four pull-down menus—File, Window, SLAM, and Help—as shown in Figure 2.2.

Figure 2.2
SLAM Toolbar

RAND *TR433-2.2*

Clicking on the SLAM pull-down menu brings up nine options, as shown in Figure 2.3. Each of these menu options will be discussed in further detail in the following sections.

Figure 2.3
SLAM Pull-Down Menu

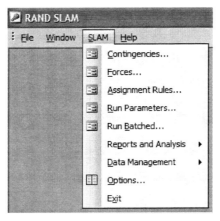

RAND *TR433-2.3*

2.2. Model Input Data

In this section, we look at the input data required before a run can be executed. Each of the subforms examined in this section must contain data in order to execute a run. We examine the existing data in the default back-end to illustrate the required data.

2.2.1. Contingencies

The first menu choice in the SLAM pull-down menu opens the Contingencies form. After clicking on the SLAM pull-down menu and selecting the *Contingencies* option, the user will see the form shown in Figure 2.4. The default back-end that we have opened contains only one contingency, named *Steady14*.

Figure 2.4
Defining a Contingency

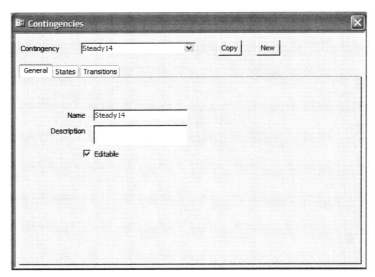

RAND *TR433-2.4*

Two types of information are required to define a contingency: states and transitions. Each of these corresponds to a separate subform on the Contingencies form. Clicking on the *States* tab opens a table with columns for state name, abbreviation, and requirements, as shown in Figure 2.5.

Figure 2.5
Defining States and Force Requirements

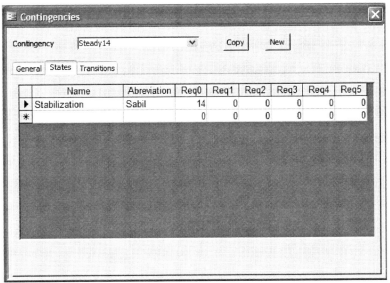

RAND *TR433-2.5*

For the contingency *Steady14*, there is only one state defined: *Stabilization*. There is also only one type of requirement defined for this state: requirement 0 (Req0) at a level of 14 units. The other types of requirements default to 0 if they are not explicitly defined by the user, as is the case in this example. In the RAND SLAM program, units are defined by the user and must be used consistently throughout. A unit can be defined as a brigade, division, individual, etc. Following *Stretched Thin*, in the analyses contained in later chapters of this report we (implicitly) define a unit as a brigade.

The last subform on the Contingencies form is the Transitions subform. This subform allows the user to define the transition probabilities to and from each state in the contingency. Since the contingency *Steady14* has only one state (*Stabilization*), the transition probability is simply equal to 100 percent (from *Stabilization* to *Stabilization*), as shown in Figure 2.6.

Figure 2.6
Defining Transition Probabilities

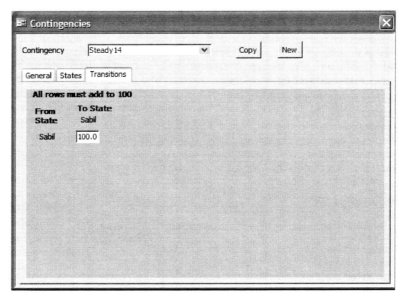

RAND *TR433-2.6*

To close the Contingencies form, the user should click the *X* in the upper right-hand corner of the form. This closes the form and saves all changes that were made while the form was open (this is true for all of the forms in the SLAM program).

2.2.2. Forces

The second menu choice in the SLAM pull-down menu opens the Forces form. This form allows the user to create and define various types of forces.

Clicking on the *Forces* option brings up the General subform, as shown in Figure 2.7. This subform allows the user to name, describe, and specify the maximum number of periods home and deployed for a force. In the back-end database SLAM1be.mdb, there is only one type of force defined: *Active*. For this force, the maximum number of periods home and deployed is set to 24. The maximum number of periods home/away defines an absorbing state. For example, if in period *n* of a model run a unit has been at home for 24 periods and is assigned to be home in period *n+1*, then this unit will be assigned to the state of being home 24 periods in period *n+1*. It is important to distinguish the time periods that define the state of a unit and the time period that defines the iteration of a model run. The model simulates *N* time periods, and during each of these time periods, any given unit will be either home or deployed for some number of periods. Time periods are defined by the user and must be used consistently throughout the program. In the analyses contained later in this report, we define a period to be a calendar quarter.

Figure 2.7
Defining a Force

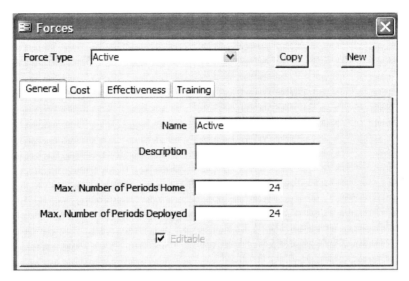

The user can select the next subform on the Forces form by clicking on the *Costs* tab. This subform allows the user to specify the per-period cost of the force while home, deployed, or training. We see from Figure 2.8 that the *Active* force is defined to have the same cost when home or deployed: 90 (training costs are ignored here). All cost parameters used in this document are calculated in terms of days per quarter and are based on a primitive cost model. Since actives are paid for every day of the month, their cost is estimated as the average number of days per month (30) times three months per quarter. Cost units are (implicitly) defined by the user and must be used consistently (days/period, dollars/period, etc.).

Figure 2.8
Defining Force Cost

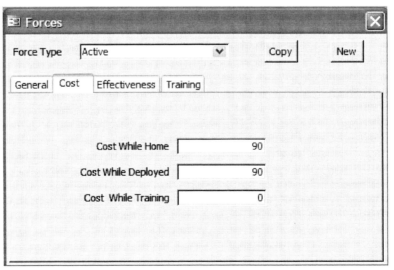

Clicking on the *Effectiveness* tab opens the subform shown in Figure 2.9. The six different kinds of effectiveness, distinguished by the number following Eff (0–5), correspond by number to the six requirement types on the States subform of the Contingency form (Req 0–5). In this example, each unit in the *Active* force is defined to contribute one unit of effectiveness to Req0 in all periods in which they are deployed (1–24), indicated by the 1 under "Eff0." In the RAND SLAM model, forces can contribute to the military requirement only when they are deployed. Eff1–5, the effectiveness for requirements 1–5, is set to 0 by default. In this example, we do not use these other requirements.

Figure 2.9
Defining Force Effectiveness

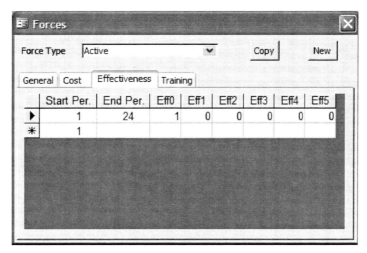

RAND TR433-2.9

Clicking on the *Training* tab opens the last subform on the Forces form. This subform, shown in Figure 2.10, allows the user to define the periods of deployment in which a force requires any of five different types of training (Tr1–5). For the *Active* force defined in the default back-end, training requirements are not considered and are therefore set to 0.

Figure 2.10
Defining Training Requirements

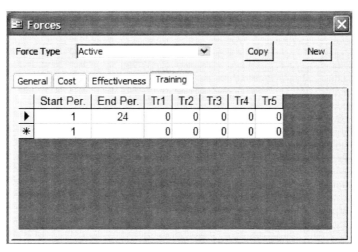

RAND TR433-2.10

2.2.3. Assignment Rules

Clicking on the *Assignment Rules* option in the SLAM pull-down menu opens the third form that must be filled in before executing a run. This form allows the user to specify the forces he would like to use and the order in which he would like to deploy these forces. The Assignment Rule form contains four subforms: General, Forces, Assignment Rule, and Advanced Assignment Rule.

The first two subforms allow the user to define the characteristics of an assignment rule. Clicking on the first tab opens the General subform, which allows the user to name and describe the assignment rule. The only assignment rule in the default back-end is called *ActivesOnly*. Clicking on the second tab opens the Forces subform, which allows the user to select which forces to include in the assignment rule. For the *ActivesOnly* assignment rule, only active forces are selected for inclusion, as shown in Figure 2.11.

Figure 2.11
Selecting Forces for an Assignment Rule

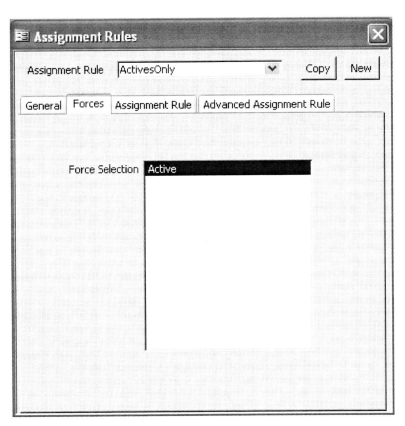

RAND *TR433-2.11*

Clicking on the *Assignment Rule* tab opens a subform that allows the user to enter the deployment ordering for groups of forces. This form creates ordinal deployment rankings (1, 2, 3, etc.) for each combination of force (active), status (home/deployed), and number of periods in status (1–24) specified. These rankings define the order in which units from each force-status-period are deployed. The ability to enter deployment rankings for groups of forces allows the user to define an assignment rule without having to enter a direct ranking for every force, status, and number of periods in status. This subform allows the user to group force-status-periods together and have the rankings automatically generated by the SLAM program. Figure 2.12 shows this subform for the *ActivesOnly* assignment rule specified in the default back-end. This assignment rule creates a deployment ranking of 1 for active units deployed three periods, 2 for active units deployed two periods, and 3 for active units deployed one period. Following this, the rankings increase incrementally for actives home 24 periods through one period (in descending order) and then continue to increment for active units deployed four through 24 periods (in ascending order). The goal of this assignment rule is to first deploy, for four periods, those active units who are at home the longest and then to decrease active time at home to one period before increasing deployment lengths. These rankings define the deployment ordering of units during a simulation run.

Figure 2.12
Defining an Assignment Rule

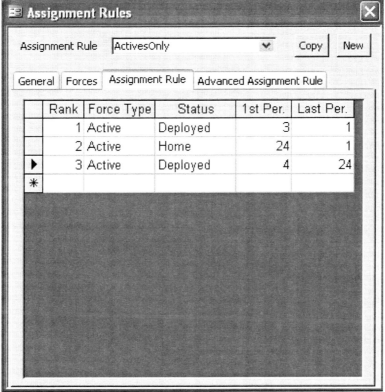

The rankings created by the SLAM program from the entries made in this subform can be seen by clicking on the *Advanced Assignment Rule* tab, as shown in Figure 2.13.

Figure 2.13
Examining an Assignment Rule

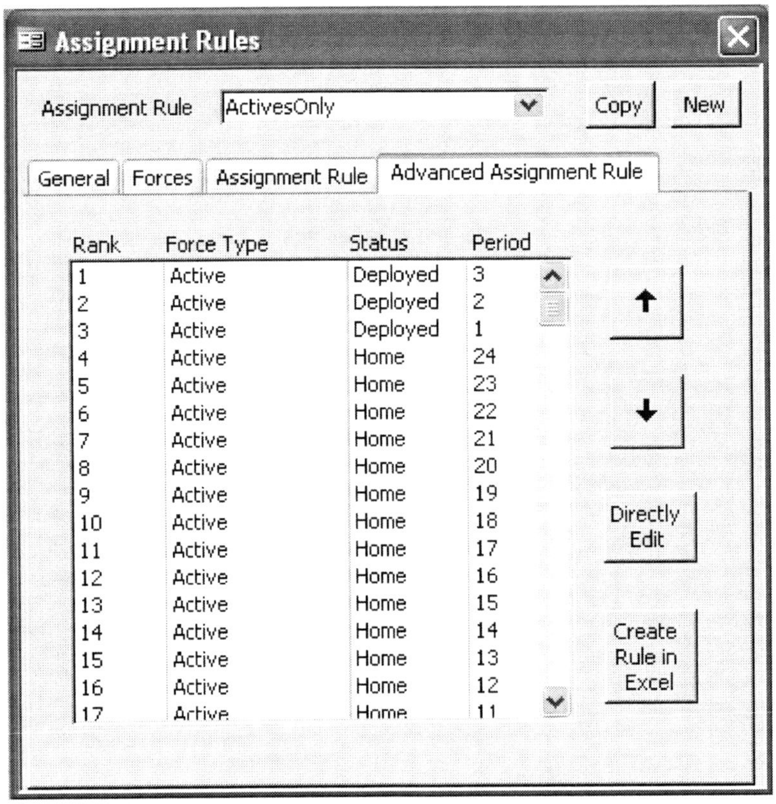
Assignment Rules
Assignment Rule
ActivesOnly
Copy
New
General
Forces
Assignment Rule
Advanced Assignment Rule
Rank Force Type Status Period
1 Active Deployed 3
2 Active Deployed 2
3 Active Deployed 1
4 Active Home 24
5 Active Home 23
6 Active Home 22
7 Active Home 21
8 Active Home 20
9 Active Home 19
10 Active Home 18
11 Active Home 17
12 Active Home 16
13 Active Home 15
14 Active Home 14
15 Active Home 13
16 Active Home 12
17 Active Home 11
Directly Edit
Create Rule in Excel

RAND *TR433-2.13*

2.2.4. Run Parameters

The final form that requires data before a simulation run can be executed is the Run Parameters form. This form brings together contingencies and assignment rules to fully define a simulation run. There are five subforms for the Run Parameters form: General, Contingencies, Assignment Rule, Initial Conditions, and Training.

Selecting the *Run Parameters* option from the SLAM pull-down menu opens the General subform on the Run Parameters form. This subform allows the user to specify characteristics for a run. Figure 2.14 shows the basic parameters for the run called *Simple Run* included in the default back-end database. This subform defines the number of periods, the start and end periods, parameters for the look-ahead periods (described in Appendix C), whether to solve the program as an integer or linear program, and the type of look-ahead methodology to use (described in Appendix C).

Figure 2.14
Defining a Run

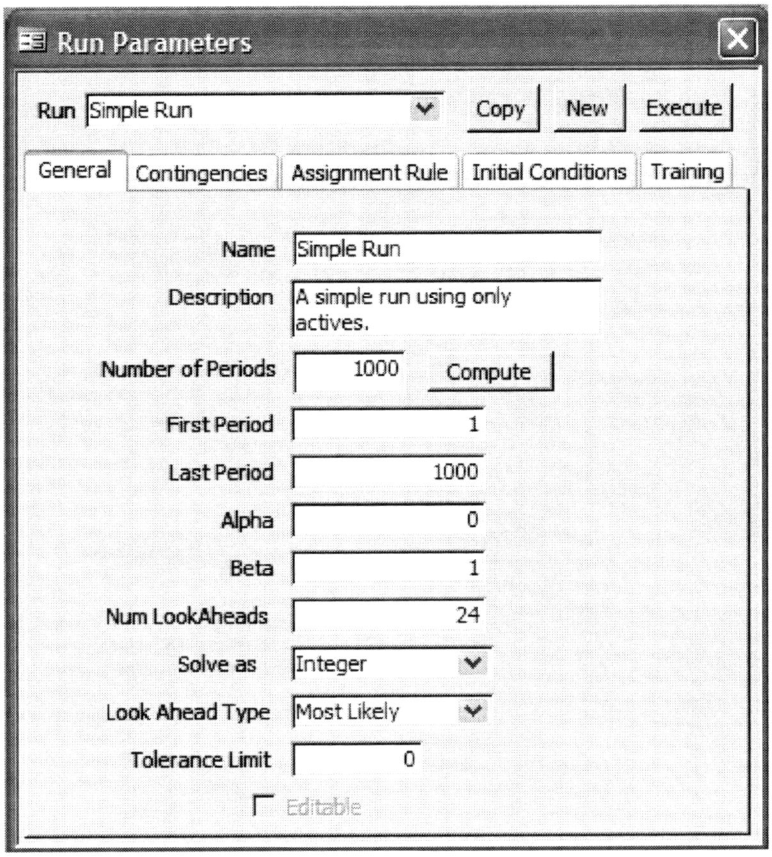

Run Parameters

Run: Simple Run   Copy   New   Execute

General | Contingencies | Assignment Rule | Initial Conditions | Training

Name	Simple Run
Description	A simple run using only actives.
Number of Periods	1000 Compute
First Period	1
Last Period	1000
Alpha	0
Beta	1
Num LookAheads	24
Solve as	Integer
Look Ahead Type	Most Likely
Tolerance Limit	0

☐ Editable

RAND *TR433-2.14*

Clicking the *Contingencies* tab on the Run Parameters form brings up a subform that shows the contingencies to be used in the selected simulation run. In the default back-end database there is only the contingency *Steady14*, which is selected for this run, as shown in Figure 2.15.

Figure 2.15
Choosing Contingencies for a Run

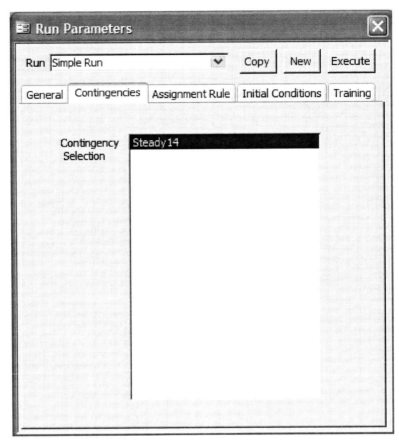

RAND TR433-2.15

Clicking on the *Assignment Rule* tab brings up a subform showing the assignment rule to be used for the selected run, along with the number of units available of each force type. In this example, the *ActivesOnly* assignment rule is selected and there are 42 active units available, as shown in Figure 2.16.

Figure 2.16
Selecting an Assignment Rule and Specifying Force Levels

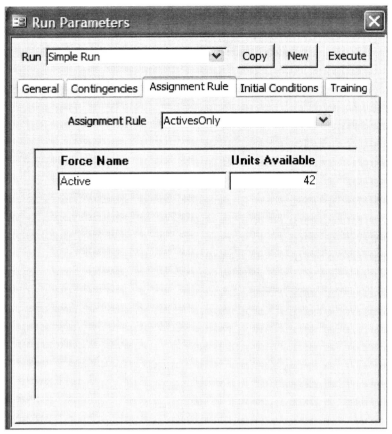

RAND *TR433-2.16*

Clicking on the *Initial Conditions* tab opens a subform that shows the initial condition for each contingency selected for this run. Since the contingency that is selected for this run (*Steady14*) has only one state (*Stabilization*), the initial condition is set to that, as shown in Figure 2.17.

Figure 2.17
Setting Initial Conditions

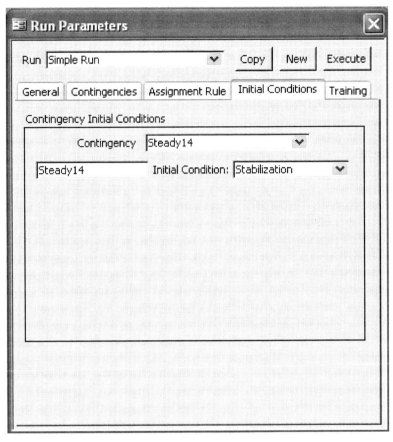

The last subform on the Run Parameters form is the Training subform. Clicking on the *Training* tab shows the training resources available for each of the five different types of training a force might require (training requirements are set on the Forces form). In this example, training constraints are not considered and are left at the default value of 0.

Before executing a run, the parameters for the run must be saved by closing the Run Parameters form after entering data. We suggest that the user close the Run Parameters form to get in the habit of doing so.

2.3. Executing a Run

We have now examined all of the SLAM input screens. The next step is to execute a run. We execute the run named *Simple Run* in the default back-end database. The user can execute a single run from the Run Parameters form. To execute *Simple Run*, the user should reopen the Run Parameters form and make sure that *Simple Run* is selected (it should be the only run in the list). After verifying this, the user should click on the execute button in the top right-hand corner of the Run Parameters form. If the run has been executed before (as is the case for the default back-end), the RAND SLAM program will open a message box asking the user if he is sure he wants to execute this run again, as shown in Figure 2.18. Click *Yes* to continue.

Figure 2.18
Execute Run Message Box

This run will take 2–3 minutes to complete. When the simulation run is completed, a message box will appear, as shown in Figure 2.19. Click *OK*.

Figure 2.19
Done with Run Message Box

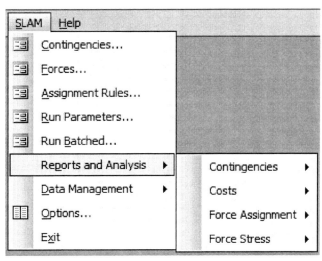

2.4. Model Output

After executing a run, there are a variety of reports and graphs that the user may wish to examine. To access them, open the SLAM pull-down menu and select the Reports and Analysis submenu, as shown in Figure 2.20.

Figure 2.20
SLAM Reports Pull-Down Menu

SLAM	Help	
▣	Contingencies...	
▣	Forces...	
▣	Assignment Rules...	
▣	Run Parameters...	
▣	Run Batched...	
	Reports and Analysis ▶	Contingencies ▶
	Data Management ▶	Costs ▶
▤	Options...	Force Assignment ▶
	Exit	Force Stress ▶

The SLAM program output is separated into four categories: Contingencies, Costs, Force Assignment, and Force Stress. In each of the following sections, we look only at the most relevant reports. Other reports are described in Appendix A.

2.4.1. Contingency Reports

The user can access the contingency reports that are available by going to the SLAM pull-down menu and clicking on the Reports and Analysis submenu and selecting the Contingencies submenu, which offers six reports, as shown in Figure 2.21.

Figure 2.21
SLAM Contingency Reports Pull-Down Menu

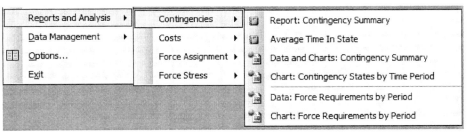

RAND *TR433-2.21*

Contingency Summary Report. Clicking on *Report: Contingency Summary* opens a Microsoft Access report, which displays the number and percentage of periods in which each contingency was in each state for each simulated run in the back-end database. Figure 2.22 shows this report for the run we just executed, *Simple Run*. The contingency summary for this run is trivial, since there is only one state in the contingency (*Stabilization*). Therefore, all periods were in the *Stabilization* state.

Figure 2.22
Contingency Summary Report

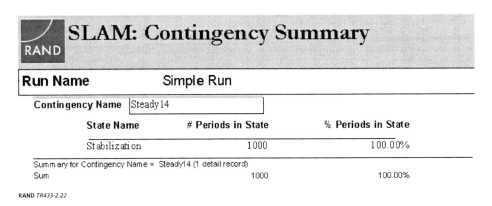

RAND *TR433-2.22*

This report can be closed by clicking the *X* in the top right corner of the window. The other five contingency reports are not relevant for this simple run but are described in Appendix A.

2.4.2. Cost Reports

Cost reports are accessible by going to the SLAM pull-down menu, clicking on the Reports and Analysis submenu, and selecting the Costs menu. This submenu lists five reports, as shown in Figure 2.23.

Figure 2.23
SLAM Cost Reports Pull-Down Menu

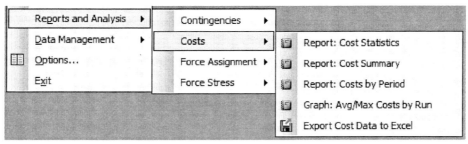

RAND *TR433-2.23*

Cost Statistics Report. Clicking on *Report: Cost Statistics* opens a Microsoft Access report, which presents the average, minimum, and maximum costs for each run. Figure 2.24 shows this report for the run we just executed.

Figure 2.24
Cost Statistics Report

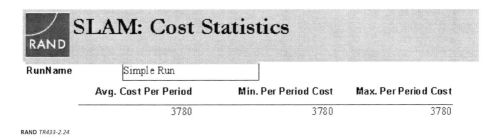

RAND *TR433-2.24*

The Cost Statistics report allows the user to easily compare the average, minimum, and maximum costs per period across runs. If the user requires more detail, we encourage him or her to explore the other cost reports (described in Appendix A). The cost statistics are trivial for *Simple Run* because the cost of active units was defined to be equal when home or deployed; therefore, the cost in each period is always the number of active units (42) multiplied by the cost/unit/period (90), which equals 3,780.

2.4.3. Force Assignment Reports

There is only one Force Assignment report, which can be used to inspect the resulting force assignment. This report is accessed by going to the SLAM pull-down menu, selecting the Reports and Analysis . . . Force Assignment submenu, and clicking on the *Data: Force Distribution by Period* option, as shown in Figure 2.25.

Figure 2.25
SLAM Force Assignment Reports Pull-Down Menu

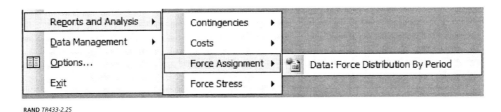

RAND TR433-2.25

Selecting this option opens a Microsoft Data Access page with a PivotTable that shows the distribution of forces in each period. This PivotTable allows the user to verify whether forces are being deployed in the order in which the user expects them to be. Figure 2.26 shows the force assignment results for *Simple Run.* The columns in this table represent each of the periods over which the simulation was run, and the rows represent each of the possible states in which a force can be in each period. For each force, there are rows for home and deployed, with the number of rows for each defined by the maximum number of periods that were entered by the user on the Forces form. The cell values in this table represent the number of units in each force-status-period in each period of the simulation. Earlier, we defined the requirement for *Simple Run* to be 14 units in each period. In Figure 2.26, we see that 14 units are deployed in every period. We can also see from this table that the model meets this requirement by deploying active forces for four periods (quarters) and in the long run allowing units eight periods at home before being redeployed.

Figure 2.26
Force Distribution Report

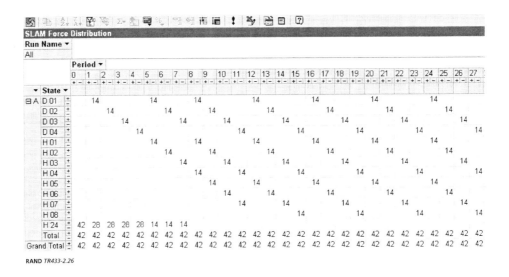

RAND TR433-2.26

2.4.4. Force Stress Reports

Force stress reports are available by going to the SLAM pull-down menu and selecting the Reports and Analysis . . . Force Stress submenu, which lists six reports, as shown in Figure 2.27.

Figure 2.27
SLAM Force Stress Report Pull-Down Menu

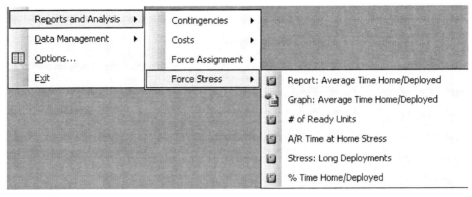

RAND *TR433-2.27*

In this section, we examine two of the six reports; the other reports are described in Appendix A.

Average Time in State Before Transition Report. Clicking on the first option in the Force Stress submenu opens a Microsoft Access report summarizing the average time at home before deployment and the average time deployed before returning home for each force type in each run. Figure 2.28 shows the results for *Simple Run.*

Figure 2.28
Average Time in State Before Transition Report

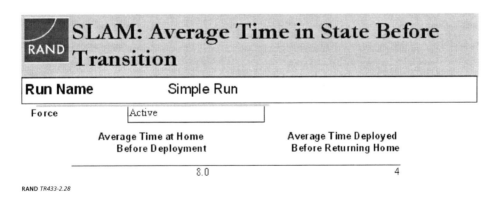

RAND *TR433-2.28*

Figure 2.28 shows that, over the course of the run we simulated, actives are at home on average eight periods between deployments and are deployed on average four periods before returning home. These metrics are very useful when there is a steady force requirement but become less meaningful when force requirements vary (described in later chapters).

Percentage Time Home/Deployed Report. The last report in the Force Stress submenu calculates the percentage of time over the course of a run that each force is home and deployed. Clicking on the *% Time Home/Deployed* menu option will open a report similar to the one shown in Figure 2.29.

Figure 2.29
Percentage Time Home/Deployed Report

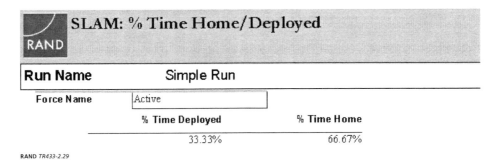

RAND *TR433-2.29*

We can see from this report that the actives in *Simple Run* spend one-third of their time deployed and two-thirds of their time at home over the course of the run. This report provides another measure of stress on the force. This metric, combined with the average time home/ deployed, provides the analyst with information useful to compare various alternatives.

2.5. Closing the SLAM Program

To close the SLAM program, the user can either click on the *X* in the top right corner of the SLAM program or click on the *Exit* option in the SLAM pull-down menu. The program will close and save all data in the active back-end database.

2.6. Summary

This chapter has provided a walk-through of the main features in the SLAM program. After following along with this chapter, the user should be familiar with the basics of using the RAND SLAM program. This chapter has walked through opening the RAND SLAM program, examining input data for an already-existing run, executing a run, and examining selected output data. The next chapters show how to create new input data and new runs, with the aim of providing useful analysis.

CHAPTER THREE

Reproducing *Stretched Thin* Using Only the Active Component

The previous chapter showed how to open, execute, and examine an existing RAND SLAM run. The next three chapters walk the user through creating new runs. These chapters demonstrate the extensibility of the RAND SLAM program using the policy challenge of supplying stabilization forces for Iraq as of late 2004 as an example. This problem has been carefully analyzed in many studies, including Congressional Budget Office (2003, 2005), O'Hanlon (2005), Davis et al. (2005), and Krepinevich (2006). Specifically, these analyses assume that the then-current Iraq environment will continue forever, and they use simple spreadsheet calculations to explore how the Army's current force structure (and incremental changes from it) could be used to satisfy the requirement. The next two chapters of this report show how to use the RAND SLAM program to specify the various policy alternatives considered in *Stretched Thin: Army Forces for Sustained Operations* (Davis et al., 2005). These chapters validate the SLAM model results against those from the *Stretched Thin* analysis. In the *Stretched Thin* analysis, force requirements are assumed to be constant over time. While the power of the SLAM model lies in its ability to simulate varying requirements, reproducing the *Stretched Thin* analysis allows us to validate the model results against another analysis that is recognized as valid.

The analysis in this chapter uses only active-duty forces to meet various operational requirements. Two metrics are used to compare these scenarios: active component time at home between deployments and the number of ready active component brigades. This chapter shows how the RAND SLAM program can reproduce these outcome measures for a variety of scenarios.

3.1. Creating a New Back-End Database

The default back-end database used in Chapter Two, SLAM1be.mdb, should be maintained in its current state so that the user can utilize it as a template in future analyses. If the user makes changes to SLAM1be.mdb and wishes to use to the original version, the original can always be copied from the SLAM distribution CD. Before proceeding with creating new input data, we create a new back-end just for this chapter. The user should reopen the SLAM program and select SLAM1be.mdb. Next, the user should click on the SLAM pull-down menu and select the Data Management submenu, then select *Create New Database*, as shown in Figure 3.1.

Figure 3.1
Data Management Pull-Down Menu

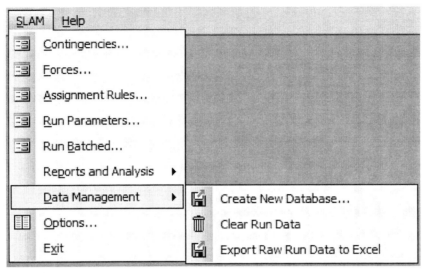

This opens the Create New Database form. This form allows the user to select which forces, contingencies, and assignment rules will be copied into a new back-end database. In this case, we choose to copy the *Active* force because we will use it again, as well as the contingency *Steady14*, as shown in Figure 3.2. Even though we will no longer use this contingency, it is generally good practice to copy at least one item from each section into the new database to avoid program errors due to missing data items. Lastly, we choose to copy the *ActivesOnly* assignment rule.

Once we have selected all of the appropriate items, we click the *Create* button. This brings up a windows prompt that asks the user to name and select a location for the new database. For the analyses used in this chapter, we created a back-end called "3-StrechedThin_ActivesOnly.mdb." Once the new database is named and saved, the new back-end database is created and ready for use. To open the new back-end, the user must first close the SLAM program, either by clicking on the *X* in the upper right corner or by selecting *Exit* from the SLAM pull-down menu.

To open the new back-end for use with this chapter, open the SLAM program (SLAM1.mdb) and click *Select Other Data File* on the SLAM start-up screen. This opens a file dialog window. Locate the file named "3_StrechedThin_ActivesOnly.mdb" and click *Open*. The new back-end database is ready for use with this chapter.

Figure 3.2
Create New Data Source Form

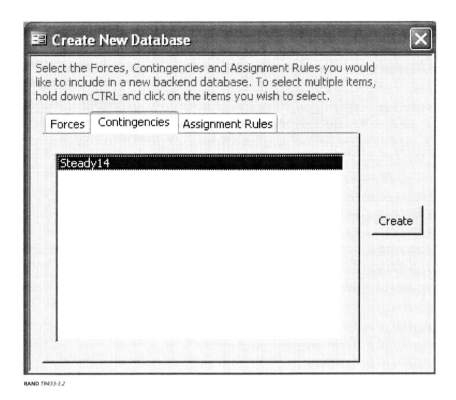

RAND *TR433-3.2*

3.2. Contingencies

The analysis in *Stretched Thin* looks at four operational requirement scenarios and three different categories of forces. The four operational requirements can be identified by the total number of brigades required: 8, 12, 16, and 20. The three categories of forces are total number of brigades, number of heavy-medium brigades, and number of infantry brigades. The number of brigades required for each type of force under each of the four scenarios is shown in Table 3.1.

Table 3.1
Force Requirement Scenarios

Scenario	Total	Heavy-Medium	Infantry
1	8	6	2
2	12	9	3
3	16	11	5
4	20	13	7

We can use the information in Table 3.1 to create the necessary contingencies in the RAND SLAM program. We need to create a contingency for each type of force under each of the four scenarios (3 force types × 4 scenarios = 12 contingencies). The contingency names are always determined by the user. In this case, we choose to name each contingency by concatenating an abbreviation of force type (Total, HeavyMed, Infantry) and the number of brigades required in the corresponding scenario. For instance, the contingency with a heavy-medium requirement of 9 will be called *HeavyMed9*.

The first step in creating the contingency *HeavyMed9* is to open the Contingencies form by selecting it from the SLAM pull-down menu. Once this form is open, the user can click on either the *New* or *Copy* buttons to create a new contingency. On this and all other forms, the *Copy* button creates a new item (in this case, a contingency) with all of the same parameters as the last selected item. The *New* button creates a new item without any prespecified parameters. To create *HeavyMed9*, we use the *New* button. After clicking this button, the user is directed to the General subform, where he or she can name and describe the contingency. The General subform for the contingency we created is shown in Figure 3.3.

Figure 3.3
Defining a Contingency

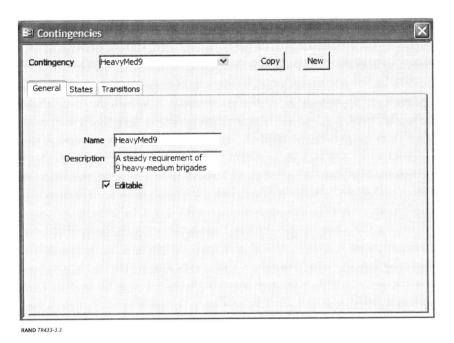

RAND *TR433-3.3*

The next step is to define the number of states for the contingency. This can be done by clicking on the *States* tab, which opens a subform displaying the table, as shown in Figure 3.4. Each row in the table represents a state for the selected contingency. Adding a new row creates a new state within the selected contingency. In this version of the SLAM program, the states can be selected from a list that includes *Peace, War,* and *Stabilization* (see Appendix A for a description of how to add states to this list). Because the analyses in *Stretched Thin* assume stable force requirements over time, this analysis requires only one state for each contingency. To enter the state, we click on the first row in the table and choose to name the state *Stabilization*. Once the state is defined, the user must specify the number of units required for each of the six requirements labeled Req0 through Req5. In all of the analyses performed in this report, there is only one total force requirement, which we choose to represent with Req0. In other situations, one may wish to include extra requirements for various types of units. This can be done by utilizing Req1 through Req5 and defining the corresponding force effectiveness on the Forces form

(Eff1 through Eff5). For *HeavyMed9*, we can look back to Table 3.1 and see that, under this scenario, 9 heavy-medium brigades are required. Again, in this report we define a unit to be a brigade. We therefore set Req0 for *Stabilization* to 9 and all of the other requirements are set to 0 by default. The completed table for this contingency is shown in Figure 3.4.

Figure 3.4
Defining the Number of States and Force Requirements

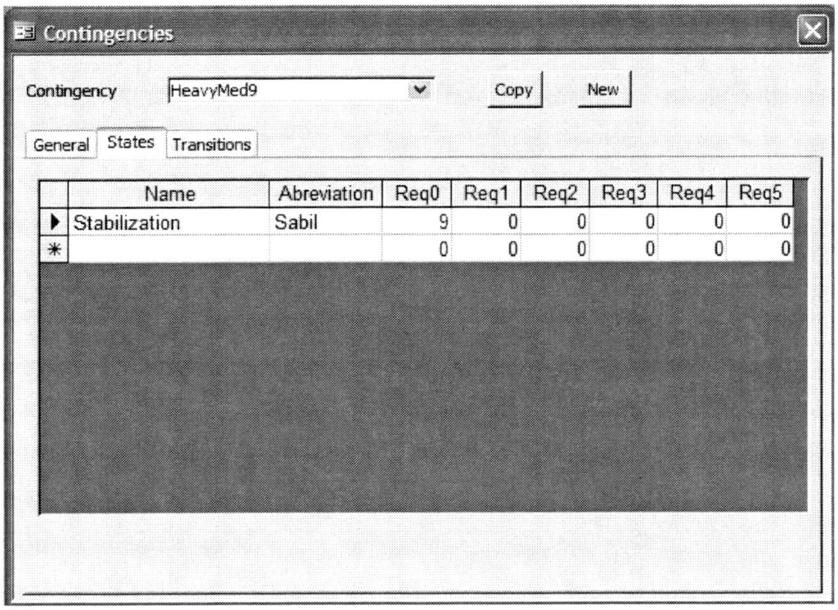

RAND *TR433-3.4*

The final step in creating this contingency is to specify the transition matrix. Each contingency defined for the analysis in this chapter has only one state, making the transitions trivial. The transition probability for each contingency is 1. This is specified on the Transitions subform as a 100 percent transition probability between *Stabilization* and *Stabilization*, as shown in Figure 3.5.

The contingency definition process described above for the contingency *HeavyMed9* was repeated for all 12 of the contingencies defined by the cells in Table 3.1. These contingencies cover all the scenarios examined in *Stretched Thin*.

Figure 3.5
Defining Transition Probabilities

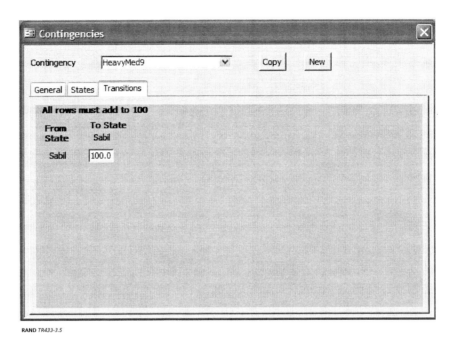

3.3. Forces

The next step is to create forces. The analysis in Chapter Two of *Stretched Thin* uses only the active component. Therefore, we need to define only one type of force: *Active*. To do this, we must open the Forces form from the SLAM pull-down menu. This form also includes *Copy* and *New* buttons, which have the same features as those on all other forms as described in Section 3.2. If we click the *New* button, we are directed to the General subform. This subform allows us to name, describe, and define the maximum number of periods home and away for each force. For all analyses contained in this report, we define a time period as a calendar quarter.

The maximum number of periods home and away defined by the user can limit the results in the Average Time Home/Deployed report. For instance, if we set the maximum number of periods at home less than the length of time a unit is actually home, we will underestimate the time at home between deployments. The longest time at home between deployments in any of the analyses in *Stretched Thin* was eight years. To replicate this analysis, we choose the maximum number of quarters home and away for the *Active* force to be 24 (six years). We do not include the full eight years, so that we can later demonstrate the effect of setting the maximum below the correct value. The general characteristics of the active component force are shown in Figure 3.6.

Figure 3.6
Defining a Force

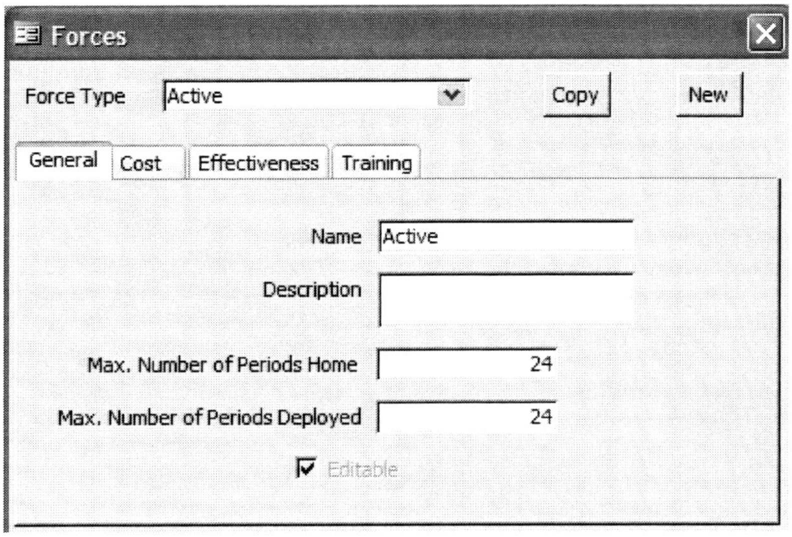

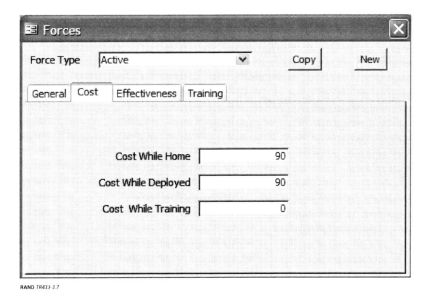

RAND *TR433-3.6*

The next step in defining a force is specifying the cost of the force while at home and deployed. This is done on the Cost subform. Using a primitive cost model, we assume that costs for active-duty forces when at home and deployed are equal. We choose to define cost in terms of the average number of days worked per quarter (90). The cost specifications are shown in Figure 3.7. We do not consider training costs in this analysis.

Figure 3.7
Defining Force Cost

RAND *TR433-3.7*

The last force characteristic that needs to be defined is effectiveness. This is done on the Effectiveness subform. Force effectiveness is defined for each period of deployment and each requirement. Force effectiveness is measured on a scale from 0 to 1, with 0 being completely ineffective and 1 being fully effective. In this analysis, we assume that active-duty forces are fully effective with regards to requirement 0 for all periods in which they are deployed. This specification is shown in Figure 3.8. The force effectiveness is associated with each unique requirement by the numeral following Eff and Req. For instance, Eff0 corresponds to Req0, and Eff1 corresponds to Req1. In this case, we are only using Req0, so we set Eff0 equal to 1 for all periods of deployment; each other force effectiveness (Eff1–5) is set to 0 by default. The requirements (Req0–5) were defined on the Contingencies States subform.

Figure 3.8
Defining Force Effectiveness

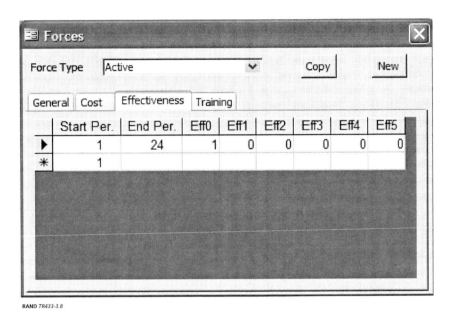

RAND TR433-3.8

The last tab on the Forces form allows the user to specify force training. However, training constraints are not considered in the *Stretched Thin* analysis and therefore are ignored here. A description of how to implement training constraints is contained in Appendix A of this document.

3.4. Assignment Rule

Before executing a run, the user must first define an assignment rule specifying the order in which to send forces. This is done by opening the Assignment Rule form from the SLAM pull-down menu. Assignment rules can be created using either the *Copy* or *New* buttons. The General subform allows the user to name and describe the assignment rule. For this analysis, we choose to create an assignment rule named *ActivesOnly*.

The next step in creating an assignment rule is selecting the forces to include. This is done by clicking on the *Forces* tab to bring up the Forces subform. Here, the user can select all of the forces to include in the assignment rule by clicking on the desired forces. In this case, we are using only *Active* forces, as shown in Figure 3.9.

Figure 3.9
Selecting Forces for an Assignment Rule

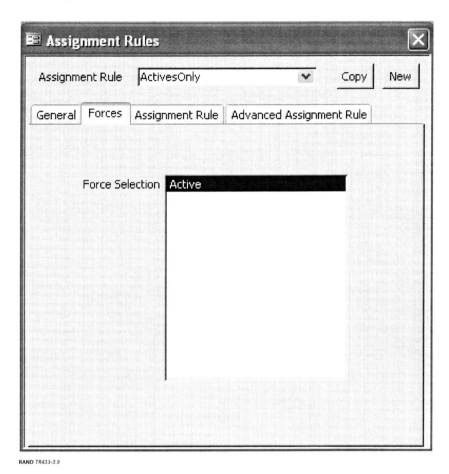

RAND *TR433-3.9*

The last step in defining an assignment rule is to specify the force ordering. This can be done in multiple ways. First, one could use the Assignment Rule subform to define how groupings of forces are ordered. This applies an ordinal ranking to the deployment of forces and is appropriate only for forces that are fully effective in all periods of deployment (i.e., not reserves). Since we are using only active-component forces and they are fully effective in all periods of deployment, this method is appropriate. (Scenarios including reserve forces require a more complicated approach to specifying assignment rules; see Chapter Four.) The table shown in Figure 3.10 allows the user to define the order in which each group of forces is deployed and have the SLAM program specify the deployment rankings (the coefficients for each of the decision variables in the linear program's objective function). For the assignment rule shown in Figure 3.10, we choose to first deploy all active-duty units already deployed three quarters (if there are any), and then to deploy those units already deployed two and one quarters by grouping all of these force-status-periods together in row 1. By setting the 1st Per. in row 1 to 3 and the Last Per. to 1, we specify that we want to start deploying forces deployed three periods first and then deploy those already deployed for two and one periods. Row 2 shows that we next choose to deploy all active-duty units at home 24 through one quarters in descending order. After deploying all forces at home, we choose to deploy all units already deployed four periods and longer in ascending order. This rule deploys all active units for one year and then reduces active time at home before deploying active units for longer than one year. The SLAM program uses the ordering in this table to create the force deployment rankings.

Figure 3.10
Defining an Assignment Rule for *ActivesOnly*

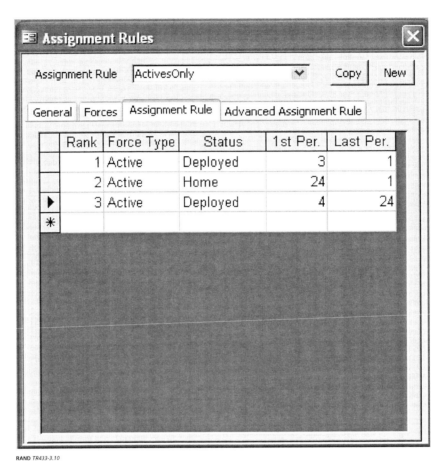

RAND TR433-3.10

It is necessary when using this table to define all possible force-period-status combinations. The user can check this by looking directly at the rankings in the Advanced Assignment Rule subform. If there is a 0 ranking for a force-status-period, this means it was not defined in the previous table.

The second method for defining force ordering is to use the Advanced Assignment Rule subform. This subform provides three different options for the user to create an assignment rule. The first option allows the user to change the ordinal ranking of the forces using arrows to move each force-status up and down, as shown in Figure 3.11. This is similar to the first method in that it gives ordinal rankings to each of the force conditions and is not appropriate for forces that are not fully effective in all periods. The user can use the first method to create a basic layout of ordering and then use the Advanced Assignment Rule subform to make incremental changes. The second option allows the user to enter assignment rule rankings directly by clicking on the *Directly Edit* button on the Advanced Assignment Rule subform. The third option creates an Excel spreadsheet to aid the user in designing an assignment rule. The spreadsheet is created by clicking on the *Create Rule in Excel* button on the Advanced Assignment Rule subform. The user must use either the second or third option when including forces that are not effective in all periods of deployment. Since the analyses in this chapter use only active forces, we ignore options two and three. More detail on using these options is provided in Chapter Four and Appendix D of this document.

Figure 3.11
Examining the *ActivesOnly* Assignment Rule

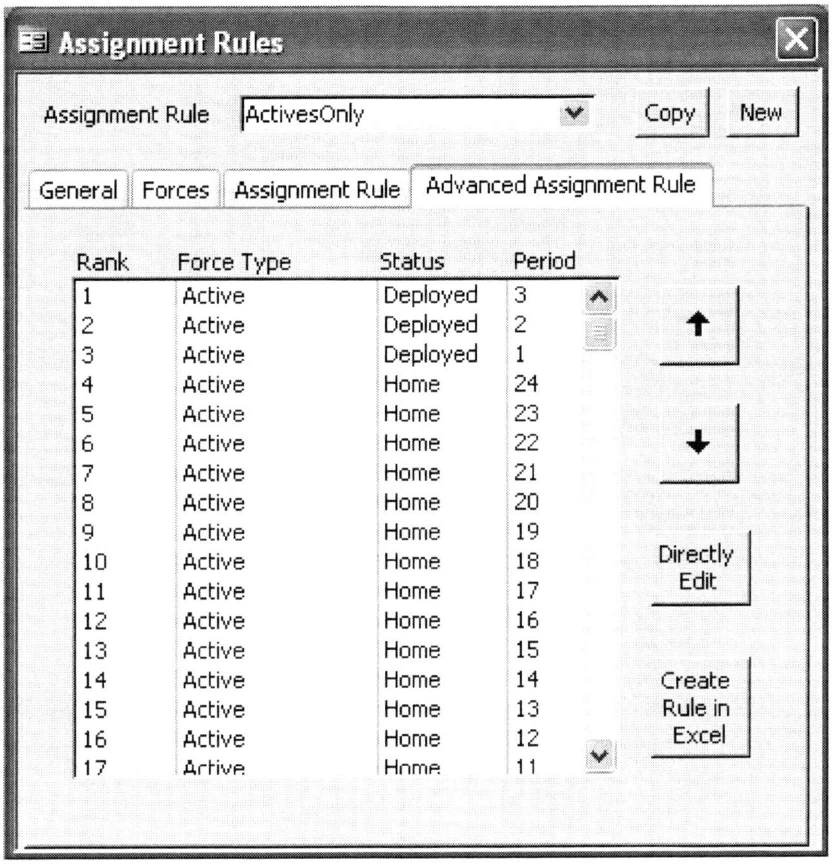

3.5. Run Parameters

The final simulation characteristics defining the run are the parameters. To reproduce the analysis in Chapter Two of *Stretched Thin*, we need 24 runs: 12 pre-transformation (baseline) and 12 post-transformation. Each run is defined by a pre- or post-transformation force structure, a type of force (infantry, heavy-medium, or total), and a force requirement scenario. The runs are named by concatenating pre/post with the type of force (Total, HeavyMed, Infantry) and number of units required. For instance, a pre-transformation, heavy-medium force under the scenario with a requirement of 9 units is named "PreHeavyMed9." We use this run as our example.

Figure 3.12
Defining a Run

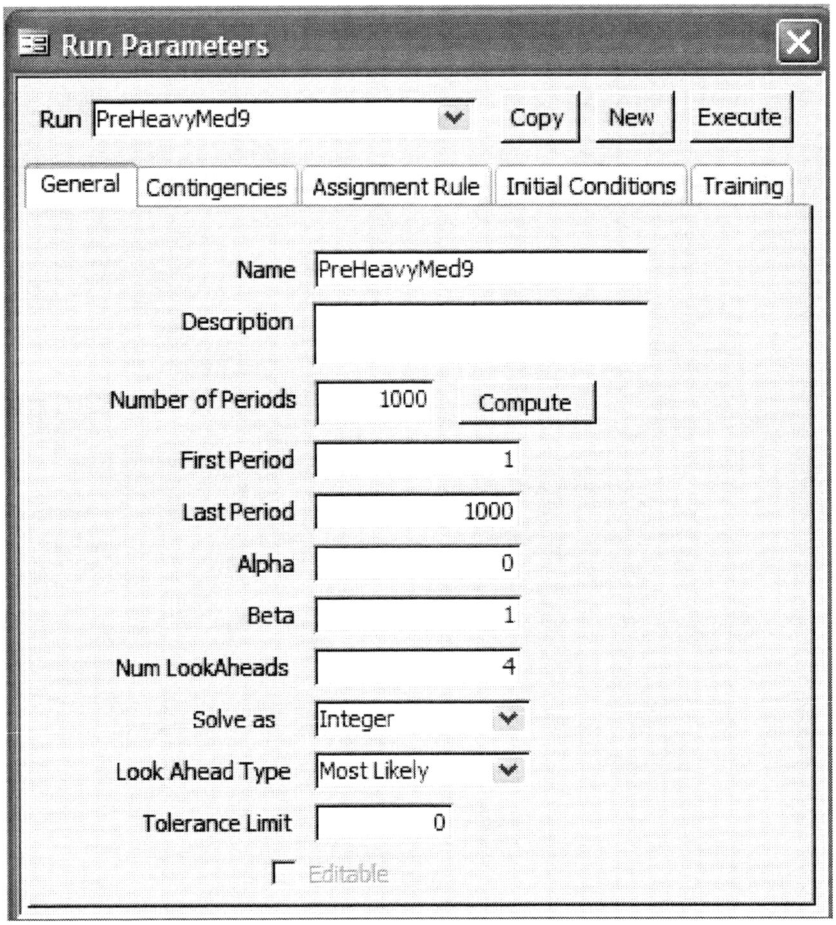

RAND *TR433-3.12*

The first step in setting up a run is to open the Run Parameters form from the SLAM pull-down menu. This opens the General subform, which allows the user to define many of the characteristics of the run. These includes name, description, period specification, parameters for look-ahead periods, and type of optimization. In this case, we choose to execute each run for 1,000 quarters (250 years), starting to collect data in period 1 and ending in period 1,000. For this analysis, we choose alpha to be 0, beta to be 1, and four look-ahead periods (the rationale for these choices is discussed in Appendix C). We also choose to simulate the run as an integer program and to use the most-likely look-ahead methodology (see Appendix C for more detail on making these choices). The parameters for the run *PreHeavyMed9* are shown in Figure 3.12.

The next step in defining a run is to specify which contingencies to include by clicking on the Contingencies subform and selecting the appropriate contingencies. In this case, the corresponding contingency is *HeavyMed9*, as shown in Figure 3.13.

Figure 3.13
Selecting Contingencies for a Run

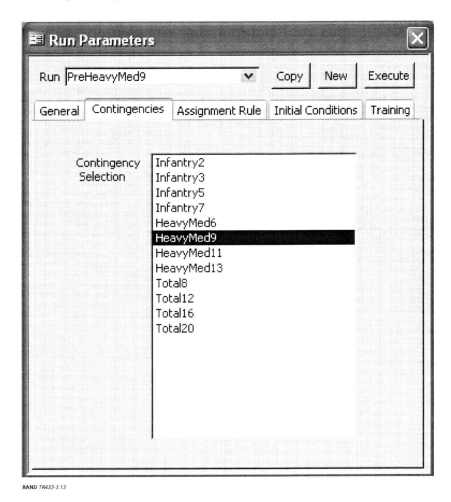

RAND *TR433-3.13*

After specifying the contingencies to include, the next step is to specify the assignment rule to use and the number of units from each force available. This is done by clicking on the Assignment Rule subform and selecting an assignment rule from the pull-down menu. Selecting the *ActivesOnly* assignment rule brings up a listing of forces included in that assignment rule with text boxes to specify the number of units available. The user must fill in these boxes for every force with the total number of units available for each force. For this example, we have two different force structures: pre-transformation and post-transformation. The numbers of available active-component units specified in Chapter Two of *Stretched Thin* are summarized in Table 3.2.

Table 3.2
Available Units by Type

Unit Type	Pre-Transformation	Post-Transformation
Heavy-Medium	21	23
Infantry	11	18
Total	32	41

Table 3.2 shows that the number of available forces for the *PreHeavyMed9* run is 21. The assignment rule specification for this run is shown in Figure 3.14.

Figure 3.14
Selecting an Assignment Rule

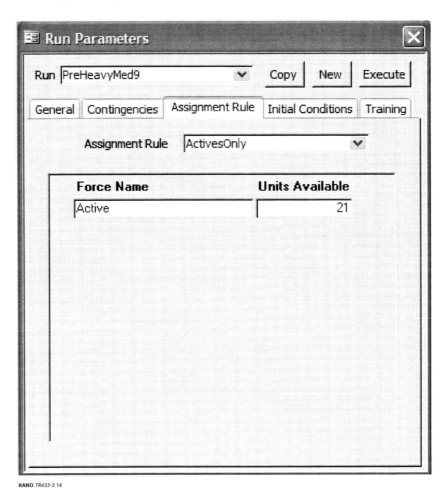

RAND *TR433-3.14*

The last piece of information needed before the user can execute a run is to set the initial conditions for each contingency on the Initial Conditions subform. This is trivial in the case of this single-contingency, single-state simulation run. The initial state must be set to *Stabilization* by clicking on the Initial Conditions subform and selecting *Stabilization*, as shown in Figure 3.15.

The process for defining the *PreHeavyMed9* run was repeated for each of the 24 runs needed for the analysis in the following sections (3 force types [HeavyMed, Infantry, Total] × 4 scenarios × 2 force structures [pre/post]).

Figure 3.15
Setting the Initial Conditions

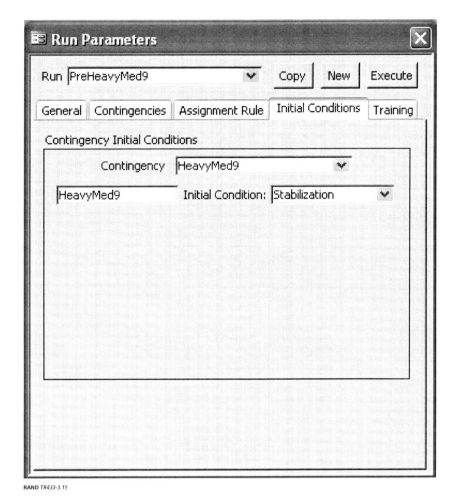

RAND *TR433-3.15*

3.6. Executing Runs

Now that all of the run parameters have been specified, we can execute the simulation runs. There are two ways to execute the runs: one-by-one or in batch mode. To execute a single run, open the Run Parameters form, select the desired run, and click the *Execute* button, as described in Chapter Two.

The Run Batched form can be used to execute multiple runs. Opening the Run Batched form brings up a listing of all the runs in the back-end database. The user can select all of the runs to be executed by changing the "Selected Run" column to "Yes" for all of the desired runs. In Figure 3.16, we choose to execute all of the pre-transformation runs. To execute the batched runs, click the *Submit Runs* button. These 12 runs executed together will take about six hours to finish. When all of the runs are finished, a pop-up window will appear.

Figure 3.16
Run Batched Form

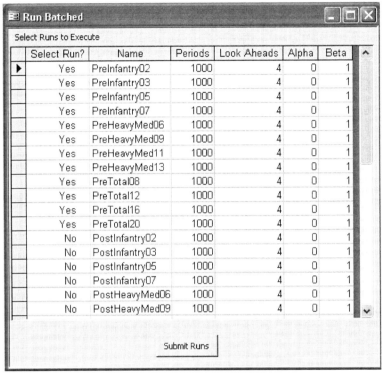

3.7. Defining Outcome Measures

As noted earlier, there are two main outcome measures utilized in Chapter Two of *Stretched Thin*. The first is average time at home between deployments. This is automatically calculated by the SLAM program and can be examined both numerically and graphically in the reports section, which is discussed in further detail in Section 3.8.

The second outcome measure is the number of ready units. After executing a set of runs, the user may wish to define this outcome measure before looking at the simulation output. The authors of *Stretched Thin* define a ready unit as any unit that has been at home for 11 months or longer. To make our specification of the simulation (in quarters) comparable to this, we assume that a ready unit is any unit that is at home for longer than 12 months (four quarters). However, we could easily define a period as a month and perform this analysis with greater resolution (with longer run times).

Unit readiness is not an important outcome measure in most SLAM analyses because of the ability to model the effects of multiple contingencies arising at the same time. For this reason, this feature is not well developed in the current version of the SLAM model. However, because unit readiness is an outcome measure used in *Stretched Thin*, we choose to show how one can calculate this. This also illustrates the flexibility of the SLAM program by showing how, with only a little knowledge of Microsoft Access, a user can define new outcome measures. Unit readiness can be defined in tblRunStress. This table can be accessed by opening any form, clicking on the "Database Window" toolbar icon (shown in Figure 3.17), and opening up the table named tblRunStress (shown in Figure 3.18).

Figure 3.17
Opening the Database Window

RAND *TR433-3.17*

Figure 3.18
Opening tblRunStress

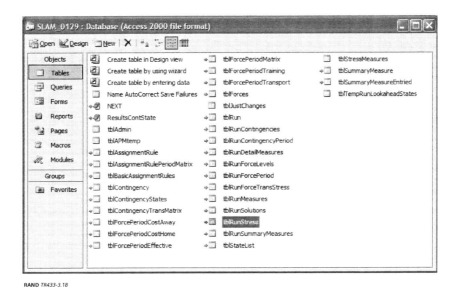

RAND *TR433-3.18*

To define a ready unit, the user must know the run sequence number (which can be obtained from tblRun) and the force sequence number (which can be obtained from tblForces). To define a ready unit, put a –1 in the Stress1, Stress2, Stress3, or Stress4 columns corresponding to the run-force-periods that the user defines as ready units. The "–1" is Microsoft Access's representation of "true." The SLAM program uses this value to calculate the total number of units in those force-status-periods marked as "True" (those defined as ready). For instance, for the run *PreInfantry8* (run sequence number=1), we can define a ready active unit (force sequence number=2) as any unit at home 5 to 24 periods (more than one year) by entering –1 in the Stress1 column for each of these periods, as shown in Figure 3.19. This same method was applied to each of the 24 runs. The results of this will be displayed in the Summary Number of Ready Units report, which is discussed later.

Figure 3.19
Defining Ready Units

RunSeqNo	ForceSeqNo	HomeAway	PeriodNum	Stress1	Stress2	Stress3	Stress4
1	2	Home	5	-1	0	0	0
1	2	Home	6	-1	0	0	0
1	2	Home	7	-1	0	0	0
1	2	Home	8	-1	0	0	0
1	2	Home	9	-1	0	0	0
1	2	Home	10	-1	0	0	0
1	2	Home	11	-1	0	0	0
1	2	Home	12	-1	0	0	0
1	2	Home	13	-1	0	0	0
1	2	Home	14	-1	0	0	0
1	2	Home	15	-1	0	0	0
1	2	Home	16	-1	0	0	0
1	2	Home	17	-1	0	0	0
1	2	Home	18	-1	0	0	0
1	2	Home	19	-1	0	0	0
1	2	Home	20	-1	0	0	0
1	2	Home	21	-1	0	0	0
1	2	Home	22	-1	0	0	0
1	2	Home	23	-1	0	0	0
1	2	Home	24	-1	0	0	0

Record: 3 of 20

RAND TR433-3.19

3.8. Results

This section compares the outcomes from the SLAM program with those from Chapter Two of *Stretched Thin*. We begin by comparing the time at home between deployments for the pre-transformation and post-transformation forces. We then compare the number of ready units under each of these force structures.

3.8.1. Average Time at Home Between Deployments

There are two ways to examine the average time at home in the SLAM program: numerically or graphically. As shown in Chapter Two, the numerical data is contained within the Average Time Home/Deployed report. This report lists both average time at home between deployments and average time deployed for each run and force in the database.

The graphical display of this data can also be accessed from the SLAM pull-down menu by selecting the Reports and Analysis . . . Force Stress submenu and then *Graph: Average Time Home/Deployed*. This opens up a Microsoft Data Access page similar to the one shown in Figure 3.20. This page contains two bar graphs—one for average time at home and one for average time deployed. Each run is plotted by name on the x-axis, and the user can choose which runs he or she wants to see displayed by clicking on the *Name* pull-down menu in the bottom left-hand corner and selecting the desired names. Figure 3.20 shows the Average Time at Home graph for only the pre-transformation heavy-medium and infantry force types. The user can also select which forces he or she wants to see displayed by clicking on the *Force* pull-down menu on the right of the graph (in this analysis, we used only active forces).

Figure 3.20
Average Time at Home Graph

Average Time at Home

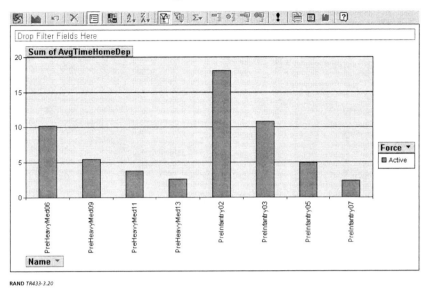

We have seen the different ways in which a user of the SLAM program can view the average time at home and deployed. We can now compare these results to those in Chapter Two of *Stretched Thin*. To do this, we extracted the numbers from the Average Time Home/ Deployed report and converted them into years. We also calculated the average time at home (in years) from *Stretched Thin* using the equation provided in the appendix of that report. This equation is:

$$\text{Average Time at Home} = \left(\frac{\text{Number of Available Brigades}}{\text{Force Requirement}} - 1\right) \times (\text{Average Deployment Length})$$

The assignment rule described earlier was designed so that the average deployment length would be four quarters, which we verified by looking at the results in the Average Time Home/ Deployed report. We use this information along with the number of available brigades and the force requirements for each run to calculate the time at home. We then compare these numbers to the results of the SLAM program. The data for the pre-transformation force is shown in Table 3.3.

Table 3.3
Results Comparison: Pre-Transformation Time at Home

Run Name	*Stretched Thin*: Active Component Time at Home Between Deployments (years)	SLAM: Active Component Time at Home Between Deployments (years)
PreInfantry02	4.5	4.5
PreInfantry03	2.7	2.7
PreInfantry05	1.2	1.2
PreInfantry07	0.6	0.6
PreHeavyMed06	2.5	2.5
PreHeavyMed09	1.3	1.4
PreHeavyMed11	0.9	0.9
PreHeavyMed13	0.6	0.7
PreTotal08	3.0	3.0
PreTotal12	1.7	1.7
PreTotal16	1.0	1.0
PreTotal20	0.6	0.6

Examining the results above, we see that the SLAM program accurately reproduces the time at home for these various pre-transformation scenarios. There are two scenarios that differ by only a tenth of a year, a difference that would likely disappear if the simulation time were increased beyond 1,000 periods and/or we ignored the initial periods.

We can also examine the same data for the post-transformation force structure. The results were derived using the same method as for the pre-transformation force and are shown in Table 3.4.

Table 3.4
Results Comparison: Post-Transformation Time at Home

Run Name	*Stretched Thin*: Active Component Time at Home Between Deployments (years)	SLAM: Active Component Time at Home Between Deployments (years)
PostInfantry02	8.0	6.0
PostInfantry03	5.0	5.3
PostInfantry05	2.6	2.8
PostInfantry07	1.6	1.8
PostHeavyMed06	2.8	2.7
PostHeavyMed09	1.6	1.5
PostHeavyMed11	1.1	1.0
PostHeavyMed13	0.8	0.7
PostTotal08	4.1	4.2
PostTotal12	2.4	2.5
PostTotal16	1.6	1.6
PostTotal20	1.1	1.1

The SLAM results shown in Table 3.4 are generally consistent with the results from *Stretched Thin*. However, there is one aberration that is worthy of discussion. For the *PostInfantry02* run, there is a requirement of 2 infantry brigades and a supply of 18 infantry brigades. Using the equation from *Stretched Thin*, we calculate the time at home between deployments to be 8 years. However, the SLAM program calculates the time at home to be only 6 years. This is because we earlier specified the maximum number of periods at home to be 24 quarters (6 years). Therefore, 6 years is an "absorbing state" for all units at home longer than this; i.e., all dwell times longer than 6 years are treated as exactly 6 years. This makes the program incapable of calculating an average time at home longer than 6 years. This can be remedied by increasing the maximum number of periods at home. In this case, a good choice would be 36 quarters, which would easily capture the expected average of 32 quarters for this run. With this exception, the SLAM results are very similar to the *Stretched Thin* results and would be even more accurate if the simulations were run for longer than 1,000 periods (with even longer run times).

3.8.2. Number of Ready Units

In the RAND SLAM program, the number of ready units is calculated based upon the definitions described earlier. The results are contained within the Microsoft Access report named "Summary Number of Ready Units." For each run, this report contains the total number of ready units across all periods, the minimum and maximum number of ready units across all periods, and the average number of ready units in any given period.

In this analysis, we are interested in the average number of ready units in any given period because this is the metric used in *Stretched Thin*. For each run, we extracted this data from the number of ready units report and compared this with the calculated equivalent from *Stretched Thin*. The equation used to calculate the number of ready units in *Stretched Thin* is:

$$\text{No. of Ready Units} = (\text{No. of Available Brigades}) \times \left(\frac{(\text{Average Time at Home}) - (\text{Preparation Time})}{(\text{Average Time at Home}) + (\text{Deployment Length})} \right)$$

The number of available brigades and average time at home vary by scenario. The number of available brigades is defined by the scenario, and the average time at home is calculated using the equation in Section 3.8.1. Both deployment length and preparation time is one year for all scenarios. If average time at home is less than one year, the equation above results in a negative number of ready units. Since this is not possible, all of these occurrences are replaced with zeroes and shown in italics in Table 3.5, which compares the number of ready units calculated in *Stretched Thin* with those calculated by the SLAM program for all the pre-transformation scenarios.

Table 3.5 shows that the results from the SLAM program are almost identical to the steady-state results in *Stretched Thin* for the pre-transformation force structure. We observed similar results for the post-transformation force structure.

Overall, the number of ready units determined by the SLAM program is almost exactly the same as those calculated in *Stretched Thin*. Any differences that do occur come about as a result of the predefined maximum number of periods at home as well as the limited duration of the simulation runs. We expect that the SLAM results would converge to exactly match the *Stretched Thin* results if each scenario were run for longer than 1,000 quarters and the maxi-

Table 3.5
Results Comparison: Pre-Transformation Number of Ready Units

Run Name	*Stretched Thin*: Active Component Number of Ready Brigades	SLAM: Active Component Number of Ready Brigades
PreInfantry02	7.0	7.0
PreInfantry03	5.0	5.0
PreInfantry05	1.0	1.0
PreInfantry07	0	0.0
PreHeavyMed06	9.0	9.0
PreHeavyMed09	3.0	3.0
PreHeavyMed11	0	0.0
PreHeavyMed13	0	0.0
PreTotal08	16.0	16.0
PreTotal12	8.0	8.0
PreTotal16	0.0	0.1
PreTotal20	0	0.0

mum number of quarters home/deployed were increased. In general, following the law of large numbers, the SLAM results will converge to the steady-state results as the number of periods is increased (Hillier and Lieberman, 2005).

3.9. Summary

This chapter has shown how to use the RAND SLAM program to reproduce the analyses performed in *Stretched Thin* using only active forces. We have seen that the SLAM results can be limited by some of the parameter choices, such as maximum time home/deployed and the number of periods. These choices must be made carefully for each run and are discussed further in the appendixes.

This chapter demonstrates only a few of the capabilities of the RAND SLAM program. The analyses contained in this chapter can be easily done in a spreadsheet. The following chapters build on these capabilities and begin to show the unique features of the RAND SLAM program.

CHAPTER FOUR

Reproducing *Stretched Thin* Using Both Active and Reserve Components

This chapter shows how to create and use forces that are not effective in all periods of deployment (such as reserves) by reproducing results from Chapter Three in *Stretched Thin* using the RAND SLAM program. The analyses in this chapter use both active-duty and reserve forces to meet a single operational requirement. The requirement utilized in this chapter is identical to Scenario Three in Chapter Three of *Stretched Thin*, in which a total of 16 brigades is required (11 heavy-medium and 5 infantry).

The first analysis in this chapter shows the effect of adding different numbers of reserve units into the rotation on active component time at home. The second analysis shows the effect of changing the reserve mobilization policies (1 year in 6, 1 year in 5, etc.) on active component time at home. The final analysis shows the effect of adding active units to the force structure on active component time at home.

4.1. Creating a New Back-End Database

The back-end database used in Chapter Three became very large (339 Megabytes [MB]) once all of the runs were executed. Because queries are run on all data contained in a back-end, it is wise to create a new back-end just for this section that copies only the simulation characteristics we will use again. This will save time by avoiding duplication of the calculations from Chapter Three when examining the reports in this chapter. This can be accomplished by clicking on the SLAM pull-down menu, selecting the Data Management submenu, and then selecting *Create New Database* and following the same process described in Section 3.1. In this case, we choose to copy the *Active* force because we will use it again, as well as all of the contingencies related to scenario 16, as shown in Figure 4.1.

Figure 4.1
Selecting Contingencies for a New Back-End Database

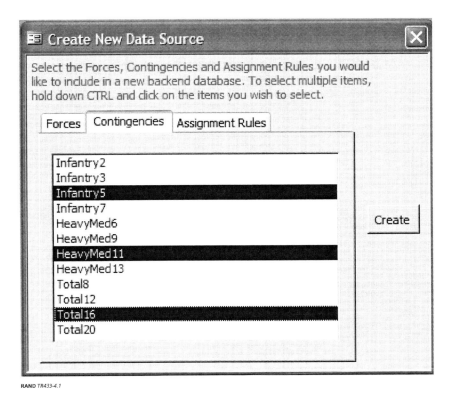

RAND *TR433-4.1*

Lastly, we choose to copy the *ActivesOnly* assignment rule. After making these selections, we click on the *Create* button and type in "4-StretchedThin_ActivesReserves.mdb" as the name of the new back-end. After closing and reopening the SLAM program, we open the new back-end by clicking on the *Select Other Data File* button on the SLAM start-up menu and selecting the new back-end file from the Windows file menu.

4.2. Contingencies

All of the contingencies needed for the analyses in this section were created in Chapter Three and copied into the back-end database that is used in this chapter.

4.3. Forces: Creating Reserve Forces

In this section, we create a base set of reserve forces that require 2 quarters of preparation time before they contribute to the operational requirement. We create this brand-new force by opening the Forces form from the SLAM pull-down menu and clicking on the *New* button. We choose to name the force *ReserveIneff2* to represent the fact that these reserves require 2 quarters (6 months) of preparation time. We choose to set the maximum number of periods home and away for the reserves to be 24. We choose this because the reserve forces in the base case scenario work on 6-year (24-quarter) rotations. Also, since reserve time at home is not an outcome measure used in the *Stretched Thin* analysis, we do not need to make the maximum lengths longer than 24.

To make meaningful comparisons, we define the cost of the reserve forces in the same way we define the cost for the active forces—number of days worked per quarter. For reservists, we assume their cost at home is 15 days per quarter (roughly 60 days per year, on a schedule of one weekend per month with double pay and two weeks per year with normal pay). We also assume the cost of deployed reservists is equivalent to that of an active-duty unit (90 days per quarter). This is a primitive cost model that considers only direct personnel costs. It does not consider cost differences related to health or retirement benefits or allowances. Substantive results are only as good as this cost model, which is poor.

Besides cost, the major difference between actives and reserves is effectiveness. In the base case used here, reservists require six months of train-up time. We implement this by making reservists ineffective for the first two quarters in which they are deployed, as shown in Figure 4.2.

Figure 4.2
Defining Effectiveness for Reserves

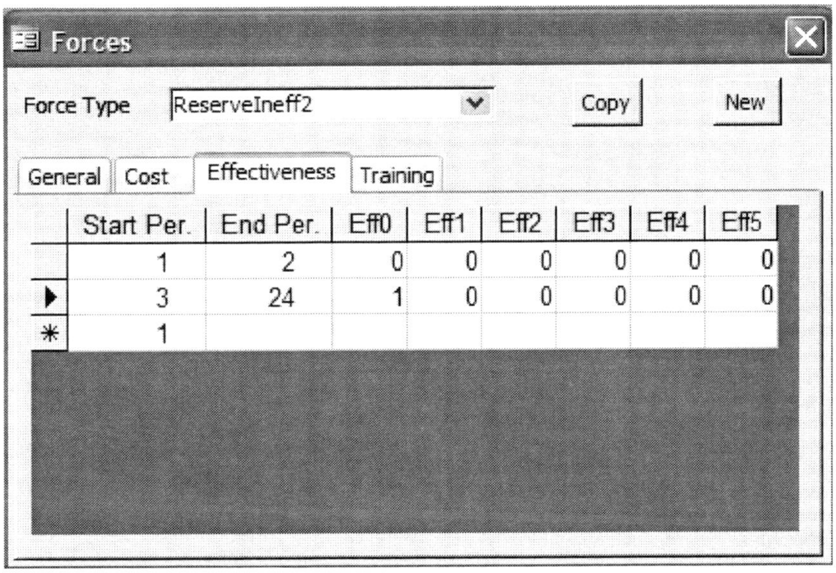

RAND *TR433-4.2*

For this analysis, we also made a change to the characteristics of the active forces. We changed the maximum number of periods home and away for active forces from 24 to 12 quarters. This reduces the run time and is appropriate given that the results in Chapter Three of *Stretched Thin* show that, for all of the scenarios being examined, active component time at home never exceeds three years (12 quarters).

4.4. Assignment Rule

The assignment rule used in the previous two chapters was simple. We ranked each force-status-period from 1 to 48 in the order in which we wanted them to be deployed. However, this approach works only for forces that are fully effective in all periods. For forces that are not fully effective in all periods (such as reserves), the user must calculate the rankings based on a heuristic and then enter them directly into the Assignment Rule form.

For the analyses in this and later chapters, we created an assignment rule that first deploys active units for up to 1 year and reduces time at home between deployments to no less than 2 years. After using all these forces, the rule deploys reserve units for deployments of up to 1 year (6-month train-up plus 6-month overseas deployment), with dwell times of no less than 5 years. Following this, the rule reduces active component time at home until the requirement is reached. We called this assignment rule *BaseActiveReserve*, and it includes both *Active* and *ReserveIneff2* forces.

When designing assignment rules involving forces not effective in all periods of deployment, the ordinal rankings used in the previous chapter must be replaced with "disutilities" of deployment. Like the ordinal rankings, the disutilities are the coefficients for each of the decision variables in the objective function of the linear/integer program that is solved to assign forces in each period. However, we changed the term from *rankings* to *disutilities* because the disutilities will not be ordinal, as the rankings we used earlier were. The disutilities will always increase for force-status-periods that are to be deployed later, but the distance between the disutilities will generally no longer be equal to one or even be of equal magnitude.

The methodology used to determine the objective function coefficients for all of the assignment rules used in this section is described in detail in Appendix D of this document. The basic premise of the assignment rule heuristic is that we should set the parameters so that the *average disutility* of deployment per effective period increases for each successive round of force-status-periods. The average disutility per effective period is defined as the sum of the disutilities of deploying a unit up to a given deployment length divided by the number of those periods of deployment in which the force is effective. To define successive groups of units, we first need to define the deployment and cycle lengths.

As an example, we will begin by describing an assignment rule for the active force with 1-year deployment lengths and 3-year cycle lengths. Under this utilization policy, an active unit deployed for 1 year should expect to be home for 2 years before deploying again. Earlier, we set the maximum number of periods home/away for the actives to 12 periods (3 years). Therefore, we first deploy units home for 12 periods for 1 period, to which we assign a disutility of 1 (the lowest disutility, so that units in this condition are deployed first). We then assign a disutility of 2 to units home 11 periods to be deployed for 1 period.

We will now discuss how the other force-status-periods are ordered for deployment (and not how the disutilities are determined; this is described in Appendix D). We set the disutility so that units home for 12 periods are deployed for 2 periods before units home for 10 periods are deployed for 1 period. Similarly, we set the disutility so that units home for 12 periods are deployed for 3 periods before units home for 9 periods are deployed for 1 period. Lastly, we set the next disutility so that units home for 12 periods are deployed for 4 periods before units home for 8 periods are deployed for 1 period.

Following this methodology will ensure that forces deployed for X periods spend on average $12 - X$ periods at home before being redeployed (for a constant force requirement). The disutilities for these force-status-periods do not increase proportionally; instead they are set to achieve targets in average disutility per effective period. The ordering of force-status-periods and their corresponding disutilities is shown in Table 4.1.

Table 4.1
Disutilities for Actives' First Use in
BaseActiveReserve

Force-Status-Period	Disutility
Active/Home/12	1
Active/Home/11	2
Active/Home/10	5
Active/Home/9	13
Active/Home/8	38
Active/Deployed/1	4
Active/Deployed/2	12
Active/Deployed/3	37

The average disutility per effective period for the active assignment rule described above is shown in Table 4.2. The methodology described in Appendix D shows how to set the disutilities shown in Table 4.1 so that the numbers in the Active/Home/12 column are larger than the corresponding values in the first row. For instance, the disutility for Active/Deployed/1 was set to 4 so that the value in the cell containing 2.5 was greater than the value in the cell containing 2.0. Setting Active/Deployed/1 to 4 made that cell value greater than the 2.0. Similarly, the other disutilities were set so that the cell containing 5.7 became greater than 5 (by setting Active/Deployed/2 to 12) and the cell with 13.5 was greater than 13 (by setting Active/Deployed/3 to 37).

Table 4.2
Disutility per Effective Period for Actives' First Use in *BaseActiveReserve*

	Periods at Home When Deployed				
Periods Deployed	Active/ Home/12	Active/ Home/11	Active/ Home/10	Active/ Home/9	Active/ Home/8
1	1.0	2.0	5.0	13.0	38.0
2	2.5	3.0	4.5	8.5	21.0
3	5.7	6.0	7.0	9.7	18.0
4	13.5	13.8	14.5	16.5	22.8

The user can utilize the methodology described in Appendix D by opening the Assignment Rule form and clicking on the *Advanced Assignment Rule* tab. From this subform, click on the *Create Rule in Excel* button to open an Excel spreadsheet that allows the user to directly import the disutilities that are specified (see Appendix D for more detail). After clicking on the *Create Rule in Excel* button, the user will see a spreadsheet similar to the one shown in Figure 4.3.

Figure 4.3
Excel Assignment Rule Template

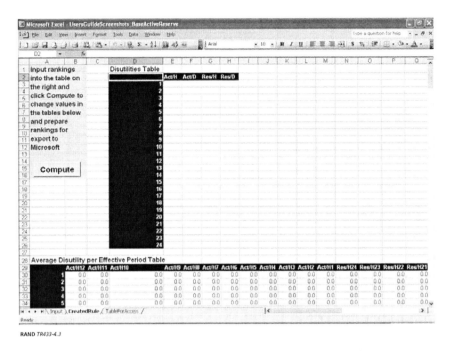

RAND *TR433-4.3*

This spreadsheet includes both a disutilities table into which the user can input the disutilities and a table that calculates the average disutility per effective period based on the user-input disutilities. When the user is finished specifying the disutilities, the disutilities will be automatically imported into the active SLAM back-end when returning to the SLAM program.

We suggest using the *Create Rule in Excel* button and the methodology described in Appendix D to specify assignment rules. However, if the user wishes to enter a set of disutilities directly, he or she may do so by clicking on the *Directly Edit* button on the Advanced Assignment Rule subform. This opens a table into which one can directly enter the disutilities in the left-hand column, as shown in Figure 4.4.

The complete specification of the assignment rule *BaseActiveReserve* is described in Appendix D. This assignment rule was created in the current back-end for use with the simulation runs in the following sections.

Figure 4.4
Directly Entering Assignment Rule Disutilities

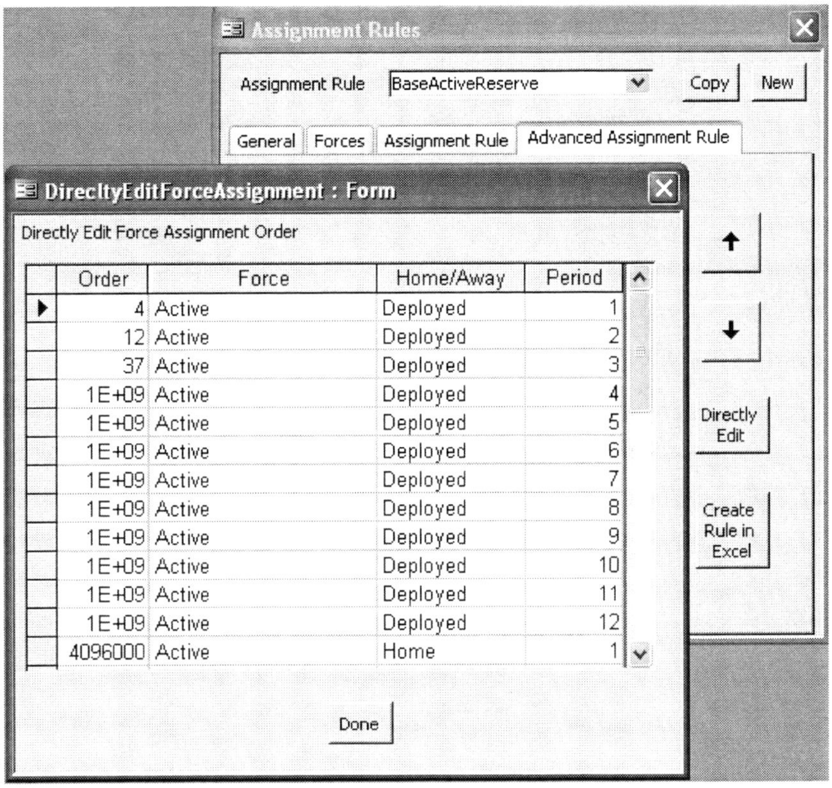

4.5. Run Parameters and Results

In Chapter Three, the run parameters for each of the executed runs were very similar. For the analyses in this chapter, which utilize both active and reserve forces, the run parameters vary depending on the specific analysis. This section describes each of these changes and shows the results of each of these changes.

4.5.1. Varying the Number of Reserve Brigades

The first analysis performed in Chapter Three of *Stretched Thin* varies the number of reserve units rotating to see how this affects active component time at home. To perform this analysis using the SLAM program, all one needs to do is define different numbers of available reserve forces for different runs. We choose to name the runs *AR_<# of rotating reserve units>*, that is, to concatenate *AR* with an underscore and the number of rotating units. For instance, the baseline case has 25 heavy-medium reserve brigades, so we call this run *AR_25*. All of the general characteristics of the runs remain the same as in Chapter Three, except for the number of look-ahead periods, which we define to be 24 (one full reserve cycle). For all of the runs in this section, we use the *HeavyMed11* contingency defined in Chapter Three. We also use the *BaseActiveReserve* assignment rule described in Section 4.4. For this analysis, we define seven different levels of reserve forces: 0, 5, 10, 11, 15, 20, and 25. The assignment rule and force levels for *AR_25* are shown in Figure 4.5.

Figure 4.5
Setting Force Levels

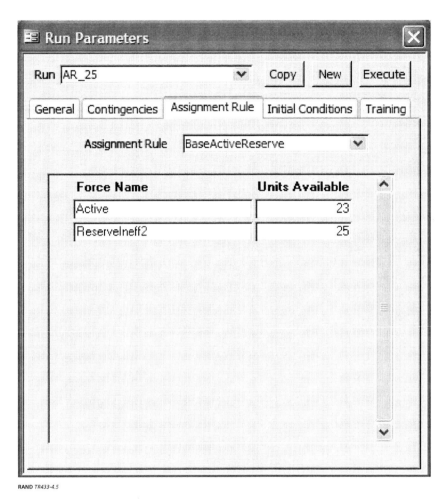

RAND *TR433-4.5*

After executing each of these seven runs for 1,000 quarters, we examined active component time at home in the Average Time Home/Deployed report. For each of the runs, we extracted the average time at home for the actives (average time deployed was 1 year in each run) and converted them into years. We also calculated the corresponding active time at home used in Chapter Three of *Stretched Thin*. To do this, we used the number of reserve component (RC) brigades (which differs by run), the fraction of time reserves are mobilized (one-sixth for all these runs), the duration of reserve mobilization (1 year), and the duration of reserve deployment overseas (0.5 year) to calculate the number of slots the reserve component could fill, using an equation from the appendix of *Stretched Thin*. For each of the runs, we subtracted this number from the requirement to determine the number of slots the active component (AC) needed to fill, and we used this number as the requirement in the time at home equation shown in Section 3.8.1. Table 4.3 compares the results calculated using the *Stretched Thin* equations with the results from the SLAM program.

Table 4.3
Effect of Adding RC Units to the Rotation on AC Time at Home

Run Name	*Stretched Thin*: Active Component Time at Home (years)	SLAM: Active Component Time at Home (years)
AR_0	1.1	1.1
AR_5	1.2	1.2
AR_10	1.3	1.3
AR_11	1.3	1.3
AR_15	1.4	1.3
AR_20	1.5	1.5
AR_25	1.6	1.5

Table 4.3 shows that the time at home calculated using the *Stretched Thin* equations and those from the SLAM program are almost identical. Any small deviations are the results of the maximum number of periods at home for the actives (12) and the number of periods (1,000 quarters). If both of these were increased, we would expect the SLAM results to converge to the *Stretched Thin* results. The main result of this analysis is that, given the assignment rule and scenario, increasing the number of rotating reserve brigades up to 25 does not increase active time at home to above the goal of two years.

4.5.2. Using the Reserve Component More Frequently

This analysis aims to determine whether using the reserve component more frequently could increase active component time at home to more than the two years desired. This analysis uses the post-transformation heavy-medium force structure (23 active units, 11 reserve units) and the heavy-medium contingency requirement of 11 units. The runs for this analysis were named by concatenating *AR* with the reserve utilization policy. For instance, the baseline utilization policy that mobilizes the reserves once every 6 years is called *AR1in6*. This run is identical to the baseline *AR_11* run from the previous analysis. Runs were created for each of the utilization policies examined in *Stretched Thin*: 1 year in 6, 1 year in 5, 1 year in 4, and 1 year in 3. The utilization frequency is defined by different assignment rules, which are described in Appendix D. Table 4.4 compares the results of utilizing the reserves more frequently from *Stretched Thin* with the results from the SLAM program.

Table 4.4
Effect of Varying RC Utilization Frequency on AC Time at Home

Run Name	*Stretched Thin*: Active Component Time at Home (years)	SLAM: Active Component Time at Home (years)
AR1in6	1.3	1.3
AR1in5	1.3	1.3
AR1in4	1.4	1.4
AR1in3	1.5	1.5

This analysis shows that increasing the mobilization frequency of the reserves still does not achieve the goal of 2 years at home for the active component. Even if the deployment frequency of reserves is increased to the level of current guidance for the actives (1 in 3), active time at home is still only 1.5 years, far from the 2-year goal.

4.5.3. Increasing the Supply of AC and RC Units

The final analysis performed in Chapter Three of *Stretched Thin* modifies the first analysis, which examined the effect of adding reserve units to the rotation on active component time at home. In this analysis, not only is the number of reserve units available variable, but so is the number of active units available. Davis et al. (2005) compare three scenarios: the base case we looked at earlier, with 23 available active component heavy-medium brigades; a scenario in which the active component adds 4 heavy-medium brigades (for a total of 27 heavy-medium brigades); and a scenario in which the active component adds 7 heavy-medium brigades (for a total of 30 heavy-medium brigades). The requirement for each of these scenarios is still 11 heavy-medium brigades. We have already examined the results from the first scenario. For the second two scenarios, we vary the number of rotating reserve units from 10 to 25, in increments of 5. Both variations in the number of active and reserve units are implemented in the SLAM program by changing the number of available units on the Run Parameters: Assignment Rule subform. The results for the second two scenarios from *Stretched Thin* and the SLAM program are shown in Table 4.5. The runs are named by concatenating *AR* with the number of reserve units, the word *Plus*, and the number of added active brigades.

Table 4.5
Effect of Increasing Supply of AC and RC Units on AC Time at Home

Run Name	*Stretched Thin*: Active Component Time at Home (years)	SLAM: Active Component Time at Home (years)
AR10Plus4A	1.7	1.6
AR15Plus4A	1.8	1.8
AR20Plus4A	1.9	1.8
AR25Plus4A	2.0	2.0
AR10Plus7A	1.9	1.9
AR15Plus7A	2.0	2.0
AR20Plus7A	2.2	2.1
AR25Plus7A	2.4	2.2

Table 4.5 shows that adding 4 active brigades increases active time at home above 2 years only if there are 25 rotating reserve brigades (the number in the pre-transformation force). However, if we add 7 active brigades, active time at home increases above 2 years if there are less than 15 rotating reserve brigades but more than 10. Again, the SLAM results are generally consistent with the results from *Stretched Thin*. We do observe that, for the runs for which active time at home increases above 2 years, the SLAM program underestimates the time at home. This is most likely the result of the maximum number of quarters at home being set to 12 (3 years), since time at home is beginning to increase toward that level. We could obtain more accurate results by increasing the maximum number of periods at home.

4.6. Summary

This chapter has shown how to create reserve forces and, analogously, any force that does not contribute to the operational requirement in all periods of deployment. We have also seen that the assignment rule specification used in Chapters Two and Three is no longer possible when we use reserve forces. Instead, we must follow the heuristic described in Appendix D to define assignment rules. The analyses from *Stretched Thin* replicated in this chapter also show that if the goal of the Army is to increase active time at home above two years, adding only active forces achieves that goal under the given scenarios. The following chapters build on this analysis by illustrating the unique features of the SLAM program. The next chapter will introduce cost as a metric and expand on the analyses from *Stretched Thin* by performing a cost-effectiveness analysis.

CHAPTER FIVE

Exploring the Cost-Effectiveness of the Reserves

In Chapters Three and Four, we examined the effect of different force structures on active component time at home and the number of ready units. These sections show that the results from the RAND SLAM program are consistent with those calculated using the spreadsheet techniques in *Stretched Thin*. A unique feature of the RAND SLAM program is that it allows the analyst to use cost as a metric when comparing these force structures. The first two sections of this chapter analyze the cost trade-off between increasing the size of the active component to reach the goal of two years at home and increasing the size of the reserve component to meet this goal. The cost estimates for active and reserve brigades used in all sections of this chapter are primitive and do not represent real costs. However, the SLAM program can be used to vary cost parameters to perform a sensitivity analysis. In the third section of this chapter, we vary the costs of the reserves to find the threshold at which they become more cost-effective than actives. In the fourth section, we change the assignment rule to allow reserves to serve for 18 months (six months' training and 12 months in theater). This analysis shows that, under the original, primitive cost parameters and an assignment rule more representative of that being followed in the global war on terror (GWOT), adding reserve forces is more cost-effective than adding active forces. In the final section, we perform the same cost-effectiveness analysis but with a goal of one year at home between deployments for the actives. Under this increased utilization of the active component and the same GWOT assignment rule, we find that adding active units is more cost-effective than adding reserve units.

This chapter is intended to illustrate how to use the RAND SLAM program to perform a cost-effectiveness analysis. The analyses in this chapter all assume that the demand for deployed forces is constant over time. Under this assumption, the results in this section can also be derived algebraically. Again, the real power of the SLAM program is the ability to perform this same type of analysis when force requirements vary over time, and this will be explored in later chapters.

We emphasize that any substantive results about cost are only as good as the cost model. The one used here is deliberately named "primitive."

5.1. Increasing the Size of Actives/Reserves

In the previous chapter, we saw that active component time at home can be increased above 2 years if 4 heavy-medium brigades are added to the active component (for a total of 27) and there are 25 rotating reserve heavy-medium units. In this section, we compare this arrangement with an alternative structure that maintains the number of active heavy-medium brigades at 23 and increases the number of reserve heavy-medium brigades to the point at which active component time at home is greater than 2 years. We used the equations from *Stretched Thin* to determine that this occurs when there are 40 heavy-medium reserve brigades, an addition of 15 reserve brigades.

To perform this analysis, we created two runs within a new back-end database. The first run (*A27R25*) uses 27 active brigades and 25 reserve brigades. The second run (*A23R40*) uses 23 active brigades and 40 reserve brigades. Both of these runs use the contingency *HeavyMed11*, in which the required number of units in all periods is 11. The assignment rule is the same basic assignment rule used in previous analyses: 1 year active deployments, 1 year reserve deployments (0.5 year effective), and reserve utilization at 1 year in 6. The first run (*A27R25*) is equivalent to the previous run named *AR25Plus4A*. The resulting active component time at home for both of these runs is shown in Table 5.1.

Table 5.1
Adding AC and RC Units to Meet AC Time at Home Goal of Two Years

Run Name	SLAM: Active Component Time at Home (years)
A23R40	2.0
A27R25	2.0

We can see that both of these configurations lead to active component time at home being equal to 2 years—the stated goal of the U.S. Department of Defense (DoD).

5.2. Cost Comparison

Now that we have seen that adding 4 active units achieves the same level of stress on the actives as adding 15 reserve units, we can compare the costs of these force configurations. In the analyses in this chapter, we again use the primitive cost parameters based on the number of days worked per quarter, as shown in Table 5.2.

Table 5.2
Cost per Quarter for Actives and Reserves

	Home	Deployed
Actives	90	90
Reserves	15	90

We can compare the costs of these two force structures by looking at the Cost Statistics report. The Cost Statistics reports for the two runs executed here are shown in Figure 5.1.

Figure 5.1
Costs of Adding 4 AC Units Versus Adding 15 RC Units

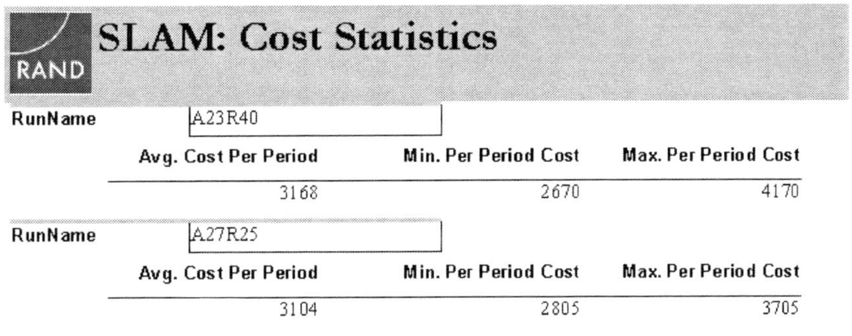

SLAM: Cost Statistics

RunName	A23R40		
	Avg. Cost Per Period	**Min. Per Period Cost**	**Max. Per Period Cost**
	3168	2670	4170

RunName	A27R25		
	Avg. Cost Per Period	**Min. Per Period Cost**	**Max. Per Period Cost**
	3104	2805	3705

RAND TR433-5.1

Figure 5.1 shows that at these levels of force utilization (actives deployed 1 year in 3, reserves mobilized 1 year in 6), adding 4 new active component brigades is less expensive than adding 15 new reserve component brigades in terms of average per-period cost (note that this does not take into account the transition costs of actually adding these brigades). With these assumptions, these results imply that, for a steady force requirement, it is cheaper to buy more active units than the corresponding number of reserve units (in terms of achieving an active time at home goal).

5.3. Reserve Cost-Effectiveness Threshold

We have seen that, under the cost parameters and assignment rule specified in the last section, buying active units is more cost-effective than buying reserve units. However, the SLAM program allows the user to change the cost of the reserves to determine the point at which the cost of buying the 15 reserve brigades is the same as buying 4 active brigades. This is a very important feature, since reliable cost estimates do not exist. This feature allows the user to perform a sensitivity analysis to determine critical cost values. To do this, we open the Forces: Cost subform and change the cost of the reserves while at home, as shown in Figure 5.2. We choose to change cost at home because deployment cost should remain the same, but with lower training or compensation costs it might be possible to reduce the cost of reserves at home.

Figure 5.2
Threshold Cost of Reserve Component

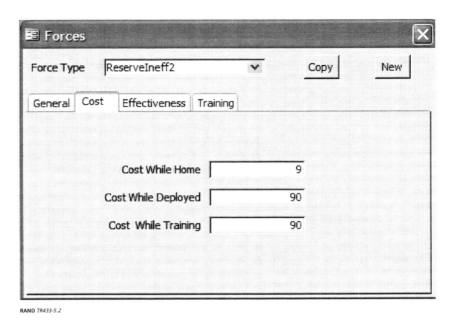

RAND *TR433-5.2*

After changing the cost on the Forces: Cost subform, it is necessary to close this form to save the parameters. Once the form is closed, the new cost data is saved and cost statistics can be examined from the SLAM pull-down menu by selecting the Reports and Analysis . . . Cost submenu and then *Cost Statistics*. After iteratively reducing the cost of the reserves at home, we found that a cost of 9 days per quarter led the average per-period cost of adding 15 reserve forces to be lower than that of adding 4 active brigades, as shown in Figure 5.3. This iterative process required decrementing the reserve home cost 6 times, which took less than 5 minutes to perform. Other analyses may require a greater number of iterations and therefore may take longer. This process may also be automated by a user with some knowledge of Microsoft Access Visual Basic for Applications (VBA).

Figure 5.3
Costs at Threshold of AC/RC Cost-Effectiveness

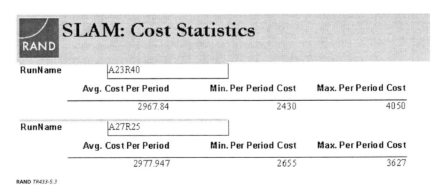

RAND *TR433-5.3*

This section shows the ease with which the RAND SLAM program can be used to perform sensitivity analyses. We have also seen the beginnings of how the RAND SLAM program can be used to analyze force size and structure changes.

5.4. Reserve Cost-Effectiveness: Global War on Terror Assignment Rule

In all previous analyses in this report, we have used an assignment rule that deploys reserves for 12 months (6 months training, 6 months in theater) every 6 years. However, in the global war on terror through 2006, reserve units were commonly deployed for 18 months (6 months in training, 12 months in theater). This section shows that, under these deployment policies and given our primitive cost model, adding reserve forces is more cost-effective than adding active forces to reach the goal of two years at home for the actives.

We begin this analysis by changing the assignment rule to reflect the alternative deployment policy of 6 quarters for reserves (4 effective). The new assignment rule is named *Active4Reserve6*. Next, we determine the reserve force level that would achieve 2 years at home for actives under this new assignment rule when 4 active units are added to the current active heavy-medium force structure (for a total of 27 active units). In the steady state, 27 active units on a 1-in-3-year rotation provide 9 effective units in any given period. In this analysis, we continue to use the *HeavyMed11* contingency with a steady requirement of 11 units in each period. Therefore, to sustain a requirement of 11 units, we need 2 effective reserve units per period to supplement the 9 effective active units. On a 1-effective-year-in-6 cycle, we need 12 reserve units (2×6). We defined a run with this new force structure and assignment rule called *A27R12_Long_AddActives* (a force structure that looks similar to the planned post-transformation reserve force plus 4 new active units).

Similarly to the previous section, we also found a force structure that would achieve 2 years at home for the active component by increasing the number of reserve units rather than active units. Twenty-three active units on a 1-effective-year-in-3 cycle provide 7.67 (23/3) effective units in any given period. Therefore, to meet the requirement of 11 units in each period, the reserves must provide 3.33 effective units in each period. On a 1-effective-year-in-6 cycle, this means there must be 20 reserve units in the force (3.33×6). We define a run with this force structure and name it *A23R20_Long_AddReserves*. We then execute each of these runs. Table 5.3 shows that both of these force structures achieve 2 years at home for the active component.

Table 5.3
Active Time at Home Under GWOT Assignment Rule

Run Name	SLAM: Active Component Time at Home (years)
A27R12_Long_AddActives	2.0
A23R20_Long_AddReserves	2.0

Next, we looked at the costs of these two force configurations under the original cost parameters (15 days/quarter for reserves at home). Table 5.4 shows that, under this deployment policy and with the primitive cost parameters in Table 5.2, adding reserve forces is more cost-effective than adding active forces to achieve the DoD goal of 2 years at home for the actives.

Table 5.4
Cost-Effectiveness of Reserves Under GWOT
Assignment Rule

Run Name	SLAM: Average Cost per Period
A27R12_Long_AddActives	2,834
A23R20_Long_AddReserves	2,743

This section has shown that simply changing the assignment rule for the reserves from 1 year in 6 to 1.5 years in 6 leads reserve units to be more cost-effective than active units. This is the same general result that can be obtained using algebraic calculations. However, algebraic calculations are limited by an assumption that wars last a long time. The SLAM model can perform these calculations for any set of assumptions about the frequency, duration, and number of wars.

5.5. Reserve Cost-Effectiveness: Using Actives More Intensively

In each of the previous analyses in this chapter, the goal was to add enough forces so that active component time at home between deployments was 2 years. However, experience in the global war on terror through 2006 has shown that active units are experiencing dwell times of 1 year or less. In this section, we perform the same cost-effectiveness analysis as in the previous sections, but at a higher requirement level (which implies shorter dwell times for active units). We will see that even under the GWOT assignment rule used in Section 5.4, when active units are deployed for 1 year in every 2, it is less expensive to add active units than reserve units to meet a specified goal of time at home for the active component.

The only change we make from the parameters used in the previous section is that we increase the force requirement from 11 to 18 brigades per period. Using the GWOT assignment rule and the base force used in the first section (23 active and 25 reserve heavy-medium brigades), we find that active time at home between deployments falls to 0.5 year when the requirement is increased to 18 brigades per period.

In this analysis, we will aim to increase active component time at home between deployments to 1 year. Table 5.5 shows that adding 5 active brigades (28 actives, 25 reserves) or adding 15 reserve brigades (23 actives, 40 reserves) to the base force achieves this goal.

Table 5.5
Adding AC and RC Units to Meet AC
Time at Home Goal of One Year

Run Name	SLAM: Active Component Time at Home (years)
A23R25	0.5
A23R40	1.0
A28R25	1.0

Next, we compare the cost of achieving the goal of 1 year at home for the active component by adding active brigades to that of adding reserve brigades. Table 5.6 shows that adding 5 active brigades is less costly than adding 15 reserve brigades to achieve the goal of 1 year at home for the actives.

Table 5.6
Cost-Effectiveness of Reserves
Under GWOT Assignment Rule and
Actives One Year at Home

Run Name	SLAM: Average Cost per Period
A23R25	2,780
A23R40	3,401
A28R25	3,345

This section shows that changing the utilization rate for the active component from 1 in 3 years to 1 in 2 years leads to active units being more cost-effective than reserve units (even under the GWOT assignment rule used in Section 5.4, which showed that reserves were more cost-effective). Again, this same general result can be obtained using algebraic calculations when one assumes that wars last a long time.

5.6. Summary

This chapter has shown how the RAND SLAM program can expand upon the analysis in *Stretched Thin* by including cost as a metric. Combining cost with other metrics, such as force stress and number of ready units, enables an analyst to perform a cost-effectiveness analysis. The ability to define different stress and readiness metrics provides the user the flexibility to perform any number of cost-effectiveness analyses using the RAND SLAM program.

This chapter illustrated how to perform a cost-effectiveness analysis when the demand for forces is constant over time. However, when the demand for forces is constant or wars are very long, these results can also be derived algebraically. The power of the RAND SLAM program is its ability to perform these calculations for any set of assumptions about the frequency, duration, and number of wars. The next chapter will perform this same type of analysis when requirements vary over time.

CHAPTER SIX

A Stochastic Perspective on *Stretched Thin*

The previous chapters showed how to use RAND SLAM in a dynamic but certain environment—the requirement for stabilization forces existed at the beginning of time and is expected to continue forever. The analysis of the previous chapters showed the circumstances under which it is optimal to use more actives or reserves.

Clearly, something is missing. That "something" is uncertainty. As their name implies, the nation maintains "reserves" for use when force requirements surge. Modeling that use of the reserves requires an environment in which force requirements vary. Perhaps one could model that environment with a deterministic (i.e., predictable, perhaps periodic) change in conflict. However, in a deterministic world, we would know when wars will start and could train up the reserves in enough time to respond to the outbreak of war. The essence of reserves as they are modeled in this report is that they are not available to surge immediately. Therefore, it seems much more realistic to view the prospect of conflict as uncertain (or stochastic); we have some sense of how often a conflict might arise, but we do not know exactly when it will arise. Furthermore, perhaps there will be more than one simultaneous conflict.

In this chapter, we show that this more-realistic stochastic environment leads to a role for reserves. This problem of uncertainty motivates the RAND SLAM program. We also show that the need to plan for the unlikely scenario of multiple simultaneous conflicts can potentially radically change the optimal force structure.

6.1. Single Stochastic Contingency

In this section, we create a single stochastic contingency and use this contingency to examine the cost-effectiveness of adding active and reserve units to the baseline force.

6.1.1. Creating a Single Stochastic Contingency

The first step in creating a new contingency is to go to the Contingencies: General subform and click on the *New* button. For the purpose of the analyses in this section, we created one contingency corresponding to one of the heavy-medium force requirement scenarios from Chapter Three. The new stochastic contingency has three states: *Peace, War,* and *Stabilization*. The contingency is defined by the number of brigades required in the *War* and *Stabilization* states, which are assumed to be the same. The requirements in the *War* and *Stabilization* states were set using the scenario described in Chapter Three, in which 11 heavy-medium brigades are required. We assume that there are no forces required in the *Peace* state.

After naming the new stochastic contingency *StochHM11*, we defined the force requirements on the States subform, as shown in Figure 6.1.

Figure 6.1
Defining States and Force Requirements for a Stochastic Contingency

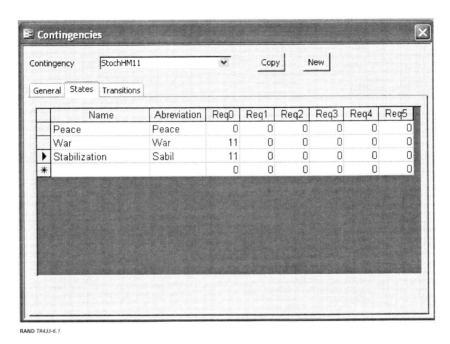

RAND *TR433-6.1*

After defining the states, the next step is defining the contingency transition matrix. We have to define transitions from and to each state listed in Figure 6.1. The transition probabilities were chosen so that, on average, a war would break out once every ten years (40 quarters). For this to be the case, the transition probability from *Peace* to *War* must be 1/40, or 0.025. We assume that one cannot transition from *Peace* to *Stabilization* because *Stabilization* can only occur after a *War*. Because each row of the contingency transition matrix must sum to 1, we can calculate the transition probability from *Peace* to *Peace* to be 39/40 (1 − (1/40)), or 0.975. The last constraint we place on the transition probabilities is that it is not possible to transition from *Stabilization* back to *War*. Although this is a dubious assumption, it allows for a simpler model (and can easily be changed and the analysis redone).

We set the remaining transition probabilities so that about 70 percent of the time will be peaceful, about 10 percent of the time there will be a war, and about 20 percent of the time there will be stabilization requirements. Appendix B describes how to calculate these steady-state probabilities. The resulting transition probabilities for this contingency are shown in Figure 6.2.

Figure 6.2
Defining a Transition Matrix for a Stochastic Contingency

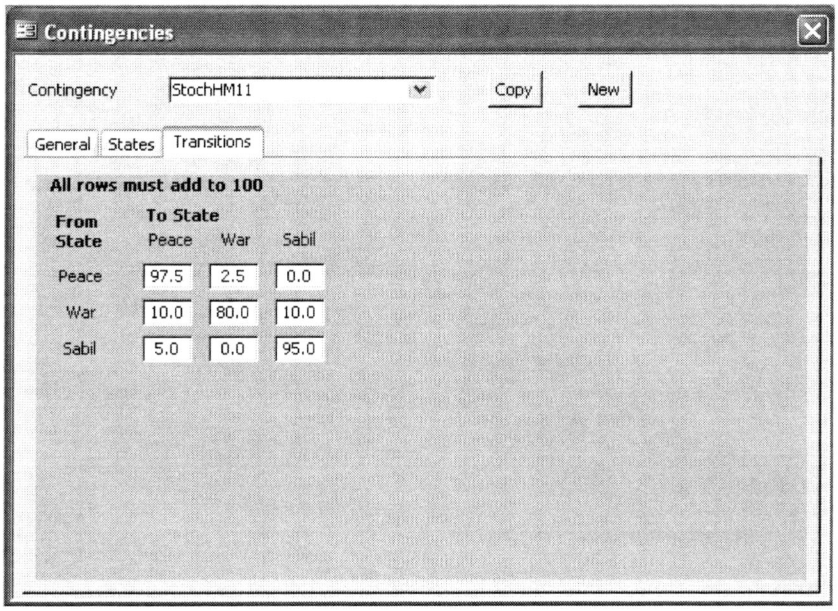

RAND *TR433-6.2*

After creating the contingency described above, we create a run named *StochHM11*, with run parameters generally the same as those in Chapter Four. We choose to return to using the *BaseActiveReserve* assignment rule, which deploys reserves 1 year in 6 (0.5 year effective). We defined the available forces using the pre-transformation heavy-medium force with 23 active units and 25 reserve units available. For each stochastic run, the user must choose an initial condition for each contingency. We choose *Peace* as the initial condition for *StochHM11*, as shown in Figure 6.3. If the simulation is run for long enough, initial conditions will not matter, since the percentage of time in each state will converge to the steady-state probabilities. However, if the user is interested in performing short-term analyses, the initial conditions are easily changed on the Run Parameters form.

Figure 6.3
Defining Initial Conditions for a Stochastic Run

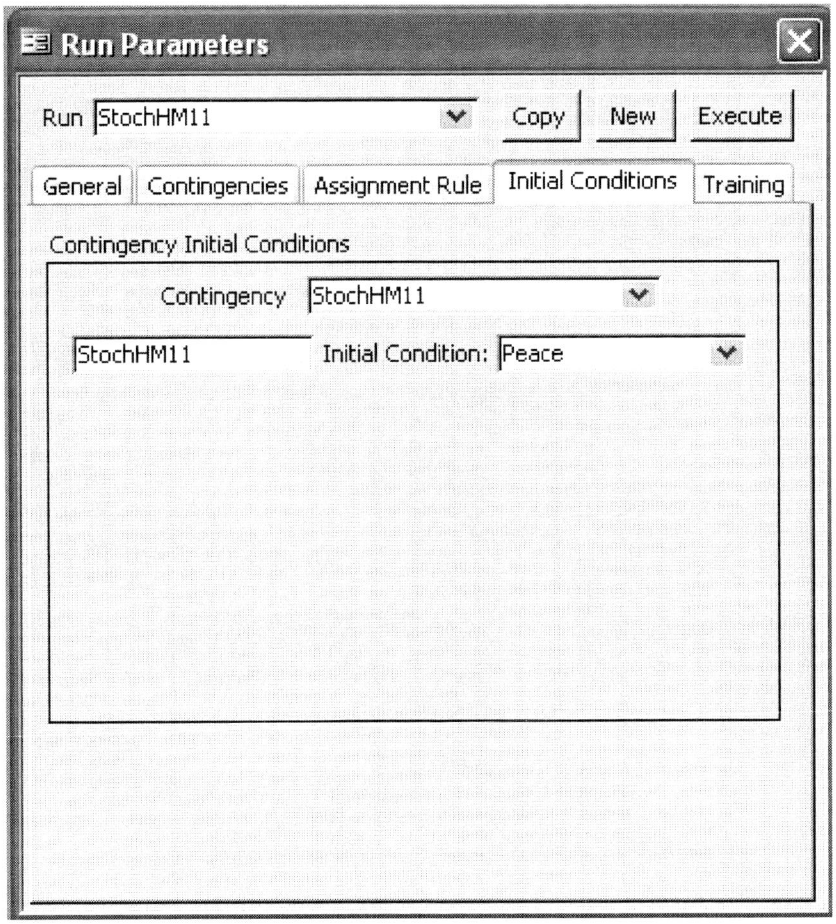

6.1.2. Single Stochastic Contingency Results

After executing the run defined above, we can examine the results. We examine outcomes for the contingencies, force stress, and cost.

Contingency Reports. After executing a run with a stochastic contingency, one of the first things the user will want to look at is a summary of contingencies. In the SLAM program, this is done by going to the SLAM pull-down menu and going to the Reports and Analysis . . . Contingencies submenu. From this submenu, we examine the two most important reports.

The first report we examine is the Contingency Summary report. For this run, a report similar to the one in Figure 6.4 appears.

Figure 6.4
Contingency Summary Report for a Stochastic Run

SLAM: Contingency Summary

Run Name	StochHM11

Contingency Name	StochHM11		
State Name	**# Periods in State**		**% Periods in State**
Peace		790	79.00%
War		110	11.00%
Stabilization		100	10.00%

Summary for Contingency Name = StochHM11 (3 detail records)
Sum   1000   100.00%

RAND TR433-6.4

Figure 6.4 shows that 79 percent of the time was peaceful, 11 percent of the time there was a war, and 10 percent of the time there were stabilization requirements. These numbers differ slightly from those calculated for the steady state discussed earlier. For all ergodic, irreducible transition matrices (there are no noncommunicating states), this will always be the case due to limited run time. However, as we increase the number of periods for which a run is simulated, the percentage of time in each state will converge to the steady-state probabilities calculated earlier. It is also worth noting here that for each run containing a contingency with the same transition probabilities and initial conditions, the contingency summary will be exactly the same. This occurs because the SLAM program uses the same random number seed (and therefore generates the same series of random numbers) for each run. The state of each contingency in each period is calculated by comparing the generated random number to the transition probabilities. Therefore, unless the transition probabilities or initial conditions are different, the same sequence of states will be generated for each contingency. This allows us to make direct comparisons of force structures, because each is fighting exactly the same series of conflicts. The program is deliberately coded in this way to allow these direct comparisons without having to simulate each run many times and then average the results (which would add significantly to run time).

The second contingency report we examined is the Average Length of States report, which calculates the average time spent in each state conditional on transitioning to the state. This report can be accessed from the Reports and Analysis . . . Contingencies submenu. For the run described in this section, the report will look similar to Figure 6.5.

Figure 6.5
Average Length of States Report for a Stochastic Run

SLAM: Average Length of States	
Run Name	StochHM11

Contingency Name	StochHM11	
State Name		**Average Time In State**
Peace		38.95
Stabilization		11.11
War		5.5

RAND TR433-6.5

The Average Length of States report shows, on average, how many periods each contingency remained in each state after initially transitioning to that state. This is important because it allows us to determine how long, on average, periods of war or stabilization lasted. Figure 6.5 shows that the duration of wars was about 1.5 years and stabilization periods lasted about 3 years on average. The results derived in this section are likely to be sensitive to the average conflict length and should be considered to be conditional on these values. As mentioned in the introduction, the results derived in this section can also be derived algebraically if one assumes that wars are long. The unique feature of the RAND SLAM program is its ability to compare the cost-effectiveness of different force structures for any set of assumptions about the number, duration, and frequency of conflicts.

Force Stress Reports. Next, we examine stress on the force. The assignment rule used for all of these runs guarantees that deployments will be no longer than 12 months for active units and no more than 6 months for reserve units (with 6 months of train-up time). Therefore, we do not consider long deployments as a stress measure in this analysis. In this section, we are interested in active time at home. In all previous chapters, we measured active time at home by calculating the average time at home before deployments (with a goal of 2 years). With a stochastic scenario, average time at home is less meaningful, because in peacetime units will generally experience very long reset times, whereas in wartime they may be deployed very frequently. Calculating average time at home would capture both of these experiences, something we would prefer to avoid. For stochastic scenarios, a better measure of stress on the force is the percentage of all deployments for which forces are home less than a given period of time when they are deployed. For the active forces, we defined three types of stressful reset times: less than 2.0 years, less than 1.5 years, and less than 1.0 year. For reserve forces, we also defined three periods of time: home less than 5.0, 4.5, and 4.0 years when deployed. The SLAM program automatically calculates these metrics for active and reserve forces. These results can be accessed from the Reports and Analysis . . . Force Stress submenu by selecting *A/R Time at Home Stress*. This brings up a report similar to the one shown in Figure 6.6.

Figure 6.6
Stress Summary Report for a Stochastic Run

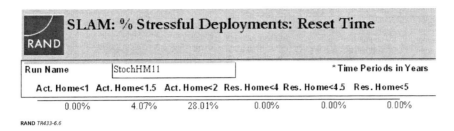

	SLAM: % Stressful Deployments: Reset Time				
Run Name	StochHM11				*Time Periods in Years
Act. Home<1	Act. Home<1.5	Act. Home<2	Res. Home<4	Res. Home<4.5	Res. Home<5
0.00%	4.07%	28.01%	0.00%	0.00%	0.00%

RAND *TR433-6.6*

The report above calculates the percentage of all deployments that occur when units are home less than the specified amount of time. The definitions of time are currently fixed in the SLAM program, but they can be easily changed with some knowledge of Microsoft Access queries. We can see that reserves never experience stressful deployments under these conditions. This is due to the fact that the assignment rule used here deploys actives and reserves on one-year cycles and then brings actives from home earlier and earlier to meet any further requirements—reserve units are never brought from home earlier.

Cost Reports. The relevant cost reports remain the same for stochastic contingencies such as those discussed in previous chapters. The main outcome of interest remains average per-period cost, which can be obtained from the Cost Statistics report in the Reports and Analysis . . . Costs submenu.

6.1.3. Force Structure and Cost-Effectiveness

In the previous section, we discussed how to set up and execute runs with stochastic contingencies. We can now modify those runs (by creating new runs with slightly different force structures) and use the outcome measures discussed in the previous section to compare the cost-effectiveness of different force structures. For this analysis, we again use the heavy-medium force requirement of 11. We choose to modify the heavy-medium force structure in a similar way to the analysis performed in Chapter Five of this report: by adding 4 heavy-medium active brigades, and by adding the necessary number of reserve brigades to achieve the same level of stress on the active force. After adding 4 active heavy-medium brigades, the force is made up of 27 active and 25 reserve heavy-medium brigades. We used this force structure and the same run parameters from the previous section with the contingency *StochHM11* (a requirement of 11 heavy-medium brigades in war and stabilization and 0 units in peacetime). We name this run *StochHM11Plus4A*. After executing this run, we calculate the stress and cost statistics from the reports discussed earlier. These results, along with those from the run *StochHM11*, are shown in Table 6.1.

Table 6.1
Stress Levels for Stochastic Runs

Run Name	Active Home Less Than 1 Year Before Deployed	Active Home Less Than 1.5 Years Before Deployed	Active Home Less Than 2 Years Before Deployed	Average Cost per Period
StochHM11	0%	4.07%	28.01%	2,502
StochHM11Plus4A	0%	0.96%	1.77%	2,858

Comparing the results from these two runs shows that adding 4 heavy-medium active brigades to the force reduces the number of active units experiencing stress (reserve stress is ignored here because reserves do not experience stressful deployments under the given assignment rule). The percentage of units being deployed with less than 1.5 years at home decreases from 4.07 percent to almost zero. The percentage being deployed with less than 2 years at home decreases from 28.01 percent to 1.77 percent. We can also see that this reduction in stress on the actives comes at a cost. The average per-period cost increases by about 360 days (4 × 90) when 4 active heavy-medium brigades are added to this force structure.

We took the level of stress defined by the *StochHM11Plus4A* run as our goal and determined the number of reserve brigades that would need to be added to the base force to achieve this same level of stress. We wanted no active units (or a very small percentage) to be deployed when home less than 1.5 years and about 2 percent of units to be deployed when home less than 2 years. After performing rough spreadsheet calculations, we determined that adding 15 reserve heavy-medium units would achieve this goal. We set up a new run for this force structure, which contains 23 active units and 40 reserve units. The results of this run, along with the results from the previous runs, are shown in Table 6.2.

Table 6.2
Stress Levels and AC/RC Cost-Effectiveness for Stochastic Runs

Run Name	Active Home Less Than 1 Year Before Deployed	Active Home Less Than 1.5 Years Before Deployed	Active Home Less Than 2 Years Before Deployed	Average Cost per Period
StochHM11	0%	4.07%	28.01%	2,502
StochHM11Plus4A	0%	0.96%	1.77%	2,858
StochHM11Plus15R	0%	0.00%	1.88%	2,757

Table 6.2 shows that adding 15 heavy-medium reserve units to the base force leads to a level of stress on the actives about equal to that of adding 4 heavy-medium active units. Looking at the costs, we see that the cost of adding 4 active units is greater than that of adding 15 reserve units to achieve the same level of force stress. This result is the exact opposite of the result found in Chapter Five under a steady force requirement and the same assignment rule. Both of these analyses used cost parameters for reserves of 15 days per quarter when at home and 90 when deployed, and 90 days per quarter for actives when both home and deployed. This analysis suggests that, when faced with varying requirements, it can be less expensive to purchase reserve brigades than it is to purchase active brigades. This occurs because reserves provide a surge capability for states of a contingency (such as war or stabilization) that have large requirements (and are sufficiently long) but are cheaper than actives during peacetime. This cost advantage of the reserves is magnified if we perform these same runs using the assignment rule for reserves examined in Chapter Five with reserves deployed 1.5 years in 6 rather than 1 year in 6. These results depend to a large extent on the underlying assumptions (length of wars, assignment rule, cost parameters, etc.), each of which can be varied by a user of the SLAM program.

6.2. Multiple Stochastic Contingencies

In Section 6.1, we created a single contingency with three states. However, in reality there is more than one threat scenario (Iraq, North Korea, etc.). Therefore, it is also important to have scenarios in which there is more than one contingency that could arise. Since much of traditional defense strategy focuses on planning for two simultaneous major regional contingencies (Larson, Orletsky, and Leuschner, 2001), this section shows how to create two stochastic contingencies and include them in the same run. We then perform a force structure cost-effectiveness analysis similar to that in Section 6.1.3 to show the extraordinary costs that must be incurred to achieve stress levels comparable to those in Section 6.1 when planning for two or more major regional contingencies (MRCs).

6.2.1. Creating and Using Multiple Contingencies

To perform the analyses in this section, we created a contingency called *StochHM9*, which has a requirement of 9 heavy-medium brigades in the *War* and *Stabilization* states and no forces required in the *Peace* state. To simplify the analysis here, we assume that both the contingencies we wish to plan for (the two-MRC requirement) are identical to the contingency called *StochHM9*. We name these new contingencies *StochHM9-1* and *StochHM9-2*.

We choose to use a requirement of 9 instead of 11 (used in the previous section) based on the ability of the assignment rule and force structure to meet the largest possible requirement. In this case, the largest possible requirement is twice the requirement in the *War* or *Stabilization* state. We found that the assignment rule could not meet a requirement of 22 units (2 × 11) without specifying a more complete rule than the one used in this document. Therefore, instead of changing the assignment rule, we simply reduced the requirement in a single *War* or *Stabilization* state to 9 (another scenario considered in *Stretched Thin*). We found that the force structure and assignment rule used earlier could meet a requirement of 18 (2 × 9). In cases in which the requirement is larger than what can be achieved with the specified assignment rule, the linear program (LP) will not solve optimally, because we placed equal "disutilities" on all force-periods beyond the rotation described in Appendix D (after completely reducing active time at home). Because we set these disutilities to be equal, the LP cannot find a solution and will return with a null result (−1s for all LP variables) for periods with this requirement after it reaches the run-time limit. The default run-time limit is 30 seconds; the user can change this value from the Options menu.

Another option available to the user to address the inability to meet military requirements is to create "phantom" forces. These are forces that do not represent real forces but will fill in when all other forces are exhausted. The usage of these forces then allows the analyst to gauge military risk—the ability to meet requirements in each period. If the user wishes to create phantom forces, he or she can utilize built-in queries to analyze military risk. To use these queries, it is suggested that the user include the word "phantom" when naming these forces. After doing this, the user can examine queries that display the percentage of periods in which the phantom forces were deployed (in the 18-%INFEASIBLE_PER query) and the average number of phantom units used in these periods (in the 18-AVGMISSING query). These queries can be accessed by opening any form, clicking on the *Database Window* icon (shown in Figure 3.17), and selecting these queries from the list of queries.

After creating the contingencies *StochHM9-1* and *StochHM9-2* in a new back-end database, we create a run called *StochHM9Multi*. The first run is exactly identical to the *StochHM11* run from the previous section, except that it has two identical contingencies with a requirement of 9 units instead of one contingency with a requirement of 11 units in the *War/Stabilization* state. This run is defined by all the same parameters used in earlier sections, except for contingencies and number of periods. This run again uses the base pre-transformation force with 23 active and 25 reserve units.

To select multiple contingencies for a run, we go to the Contingencies subform on the Run Parameters form and click on the two contingencies we want to include, as shown in Figure 6.7.

Figure 6.7
Selecting Contingencies for a Multiple-Contingency Run

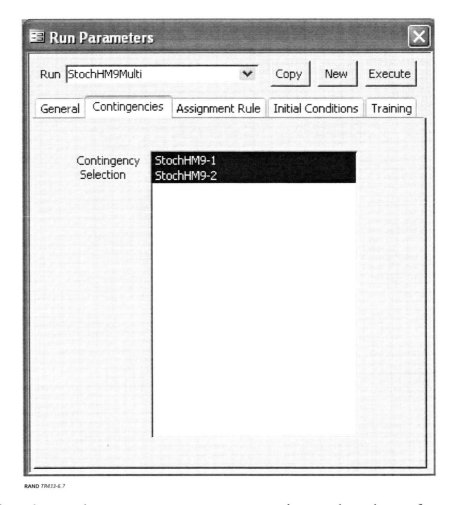

RAND *TR433-6.7*

After selecting these contingencies, we must set the initial conditions for each of them. This is done by going to the Initial Conditions subform on the Run Parameters form and, selecting each contingency separately from the Contingencies pull-down menu, setting the initial state to *Peace* for each one, as shown in Figure 6.8.

Figure 6.8
Setting Initial States for a Multiple-Contingency Run

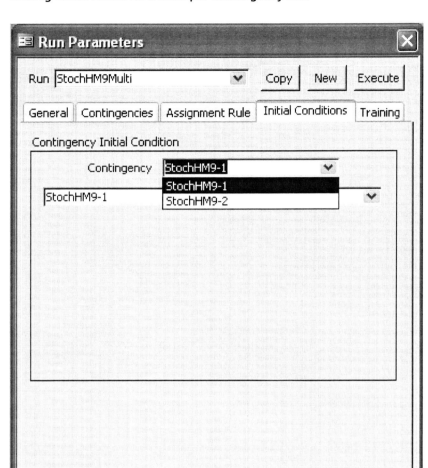

RAND *TR433-6.8*

The final change that needs to be made before we execute this run is to define the number of periods for which to execute the run. In all previous analyses in this report, we have chosen to execute runs for 1,000 periods. However, because force planning deliberately sizes the force to handle the least-likely outcomes, we increase the number of observations for stochastic runs so that we see enough of the rarest outcomes—such as two simultaneous MRCs. For the analyses we wish to perform here, we set the number of periods so that there are about 100 periods in which both contingencies are in the *War* state at the same time. More precisely, we set the number of periods so that the expected amount of time that both contingencies are in the *War* state is 100 periods.

To have the SLAM program calculate this automatically, the user must click the *Compute* button next to the *Number of Periods* input box on the Run Parameters form. Doing so opens up a form that asks the user how many periods he or she wants to see in the least-likely states, as shown in Figure 6.9.

Figure 6.9
Using SLAM to Compute the Number of Periods

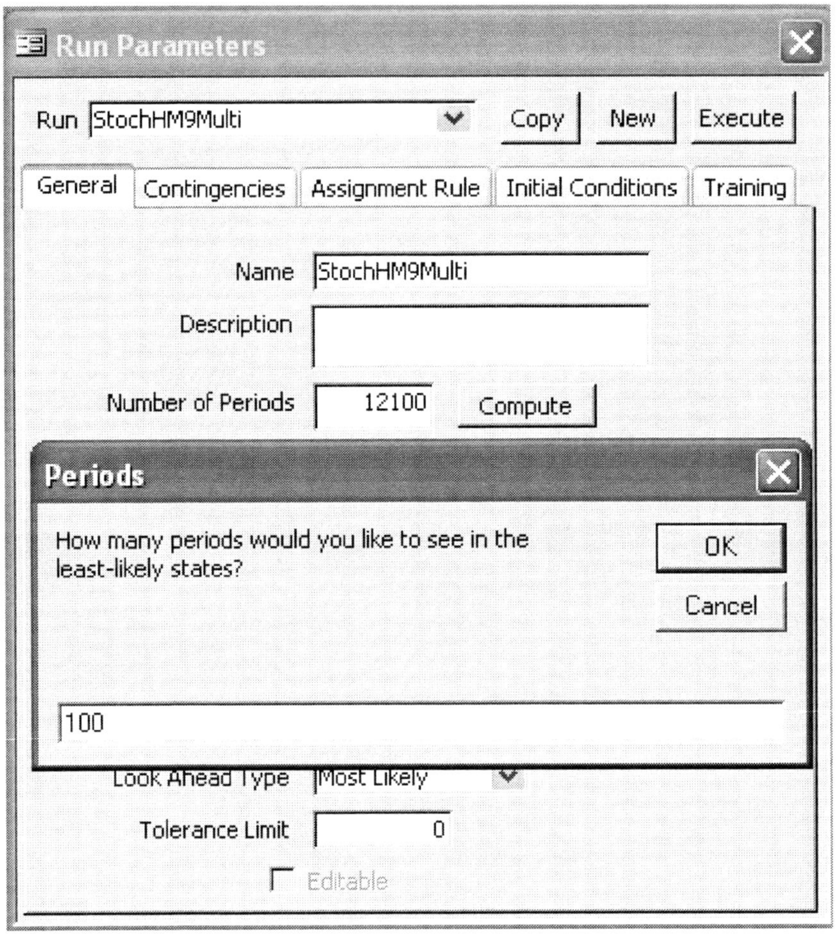

RAND *TR433-6.9*

The SLAM program calculates the number of periods necessary to see the least-likely state the number of times that the user specifies on this form. The methodology used by the SLAM program to calculate this number is described in Appendix B. The user must select contingencies on the Contingencies subform of the Run Parameters form for this to work; otherwise, an error message will appear.

For the run described in this section, the SLAM program found the minimum number of periods necessary to see the least-likely states in 100 periods to be 12,100. For the analyses presented here, we decided to execute this run for 10,000 periods to save run time. Note that the discussion in Appendix B sets the number of periods so that *at least* 100 periods will have all the least-likely states occurring at once. Here, we decided that 10,000 periods was sufficient to "see" the most unlikely state just slightly less than 100 times. After setting the number of periods to 10,000, we execute the run.

6.2.2. Multiple Stochastic Contingency Results

The first report that we examine to make sure that the run was executed correctly is the Contingency report. Figure 6.10 shows the percentage of time that each contingency was in each state over the course of the run.

Figure 6.10
Contingency Summary Report for a Multiple-Contingency Run

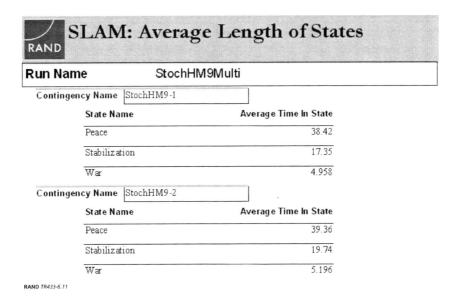

SLAM: Contingency Summary

Run Name	StochHM9Multi	

Contingency Name	StochHM9-1	
State Name	**# Periods in State**	**% Periods in State**
Peace	7382	73.82%
War	952	9.52%
Stabilization	1666	16.66%

Summary for Contingency Name = StochHM9-1 (3 detail records)

Sum	10000	100.00%

Contingency Name	StockHM9-2	
State Name	**# Periods in State**	**% Periods in State**
Peace	7135	71.35%
War	930	9.30%
Stabilization	1935	19.35%

Summary for Contingency Name = StockHM9-2 (3 detail records)

Sum	10000	100.00%

RAND *TR433-6.10*

Next, we look at the Average Length of States report to determine the average length of each state within each contingency; these results are shown in Figure 6.11.

Figure 6.11
Average Time in State Report for a Multiple-Contingency Run

SLAM: Average Length of States

Run Name	StochHM9Multi	

Contingency Name	StochHM9-1	
State Name	**Average Time In State**	
Peace	38.42	
Stabilization	17.35	
War	4.958	

Contingency Name	StochHM9-2	
State Name	**Average Time In State**	
Peace	39.36	
Stabilization	19.74	
War	5.196	

RAND *TR433-6.11*

After examining the Contingency reports, we look at the Cost and Stress reports for the multiple stochastic run. The results of these reports are summarized in Table 6.3, which shows that the multiple stochastic contingencies place a significant amount of stress on active forces.

Table 6.3
Stress and Cost Results for a Multiple-Contingency Run

Run Name	Active Home Less Than 1 Year Before Deployed	Active Home Less Than 1.5 Years Before Deployed	Active Home Less Than 2 Years Before Deployed	Average Cost per Period
StochHM9Multi	21.26%	28.35%	35.87%	2,553

6.2.3. Force Structure and Cost-Effectiveness

In this section, we perform an identical analysis to that performed in Section 6.1.3, except that in this case we use the two stochastic contingencies defined in Section 6.2.2. Here, we show that adding reserve units to the force decreases stress on the active component only up to a point, after which adding more reserve units provides very little marginal benefit (although adding active units can still significantly decrease stress).

We first create a run identical to *StochHM9Multi* but add 25 heavy-medium active brigades to the base force (for a total of 48 active units and 25 reserve units). We choose to add 25 active units because this is the number of units needed to achieve 2 years at home for the actives in a two-war scenario (the combination of states that is the most stressful). We calculate this number by determining the number of effective units per period needed from each force. In this first analysis, we have assumed there are 25 reserve units. On a 1-year-in-6 cycle (0.5 year effective), these reserves provide 2 effective units per period. Therefore, since the most stressful combination of states (*War-War*) has a requirement of 18, the actives must provide 16 effective units per period in this combination of states. To provide these forces and achieve the goal of 2 years at home, there must be 48 (16 × 3) active units in the force, an increase of 25 units.

Although it is unrealistic that 25 active units would be added to the force structure, the purpose of this analysis is to show that trying to achieve a very low stress level at all times while planning for two MRCs is very expensive. Thus, when performing this type of analysis, the trade-off between force stress and cost should be carefully considered. The SLAM program can help an analyst explore these kinds of trade-offs.

We first compare the stress and cost of the run with just the base force, with the run adding 25 active brigades. Table 6.4 summarizes these comparisons, showing that adding 25 active units reduces stress on the active component to almost zero, but that this comes at a very high cost (an increase of 2,160 days per quarter).

Table 6.4
Stochastic Stress and Cost Comparisons Adding AC Brigades

Run Name	Active Home Less Than 1 Year Before Deployed	Active Home Less Than 1.5 Years Before Deployed	Active Home Less Than 2 Years Before Deployed	Average Cost per Period
StochHM9Multi	21.26%	28.35%	35.87%	2,553
StochHM9MultiPlus25A	0.00%	0.03%	0.33%	4,713

In a similar fashion to the analysis in Section 6.1.3, we take the stress levels achieved from adding 25 active units as our goal and attempt to increase the size of the reserve component to meet these same stress levels. We begin by creating a run, adding 99 heavy-medium reserve brigades to the base force (for a total of 23 active units and 124 reserve units). We choose to add

99 reserve brigades because this number makes the time at home for the actives in the two-war scenario equivalent to that of adding 25 active heavy-medium brigades (providing actives 2 years at home between deployments). The stress and cost results of this run and the two previous runs are shown in Table 6.5.

Table 6.5
Stochastic Stress and Cost Comparisons Adding RC Brigades

Run Name	Active Home Less Than 1 Year Before Deployed	Active Home Less Than 1.5 Years Before Deployed	Active Home Less Than 2 Years Before Deployed	Average Cost per Period
StochHM9Multi	21.26%	28.35%	35.87%	2,553
StochHM9MultiPlus25A	0.00%	0.03%	0.33%	4,713
StochHM9MultiPlus99R	2.60%	6.47%	12.21%	4,269

This table shows that adding 99 reserve brigades leads to a stress level *higher* than that of adding 25 active brigades. Although the per-period cost is lower for adding 99 reserve brigades to the base force than it is for adding 25 active brigades, it does not achieve the same low level of stress on the actives. This occurs because, when a war breaks out, active component units are the only units immediately able to respond for the first six months of a conflict. Because reserves cannot replace actives in these situations, the only way to reduce stress on the actives completely is to add active units.

To test the hypothesis that adding more reserves will lead to only very small decreases in stress on the actives, we create two more runs, adding 123 and 147 reserve units to the base force. Table 6.6 shows the results of this analysis: Adding many more reserve units does very little to reduce stress on the actives. Adding more than 123 reserve units is just as expensive as adding 25 active units. Therefore, if one wishes to reduce stress on the actives below the level achieved by adding 99 reserve units, choosing to add active units instead of reserve units is a more cost-effective approach.

Table 6.6
Adding Reserves Does Little to Reduce Stress from Two MRCs

Run Name	Active Home Less Than 1 Year Before Deployed	Active Home Less Than 1.5 Years Before Deployed	Active Home Less Than 2 Years Before Deployed	Average Cost per Period
StochHM9Multi	21.26%	28.35%	35.87%	2,553
StochHM9MultiPlus25A	0.00%	0.03%	0.33%	4,713
StochHM9MultiPlus99R	2.60%	6.47%	12.21%	4,269
StochHM9MultiPlus123R	2.02%	5.06%	9.35%	4,704
StochHM9MultiPlus147R	1.96%	4.80%	8.49%	5,135

This analysis illustrates two things. First, no matter how one attempts to reduce stress on the actives, achieving a low stress level when planning for a two-MRC scenario can be very expensive. Second, beyond a certain threshold, adding reserve units does little to reduce stress on the actives, because reserve units cannot immediately respond to the outbreak of a new contingency. It is important to note that the stress measures discussed here are driven by the most stressful states (two wars) and cannot be reduced without addressing the force requirements in these states as we did above. Force changes of this magnitude are very unlikely. However, the SLAM program can be used to explore the trade-offs between force structure and cost and to help find an optimal force configuration that makes trade-offs between force stress and cost.

6.3. Summary

This chapter shows how to use the SLAM program to define stochastic threat scenarios and use them to analyze force structure decisions. The analyses in this chapter examined the role of active and reserve heavy-medium brigades in handling different types of stochastic scenarios. In Chapter Five, we found that, under a steady (non-stochastic) force requirement, it was always less expensive to add active brigades to the force instead of reserve brigades when reserves were deployed 1 year in 6 (6 months in theater). In this chapter, we found that, when faced with a single stochastic threat scenario (and using the same assignment rule), the opposite is true—it is less expensive to add reserve brigades than it is to add actives to achieve the same level of stress on the active force. For a two-MRC scenario, we showed that adding reserve units has very little impact on reducing stress on the active component below a certain level, because reserves cannot immediately respond to the outbreak of a war.

The analyses in this chapter use primitive estimates of cost and effectiveness parameters to derive simple results. A more extensive analysis would use better cost estimates to determine the sensitivity of these results.

CHAPTER SEVEN
Examining the Active-Reserve Mix

The previous chapters have used the RAND SLAM program to replicate existing analyses and to show some of the unique features of the RAND SLAM program by extending beyond these analyses. This chapter explores the problem that motivates the RAND SLAM program—determining an optimal mix of active and reserve forces. This chapter uses both deterministic and stochastic contingencies to show the effects of various changes to the active-reserve mix on cost and stress measures.

All previous chapters have focused on different types of forces (infantry, heavy-medium, etc.). This chapter looks at the structure of the total force (infantry plus heavy-medium units). This final analysis chapter brings together many of the features of the RAND SLAM program illustrated in previous chapters. However, the features of the RAND SLAM program are not limited to those shown here. With little effort, an experienced Microsoft Access user can create many more features not listed here.

7.1. Deterministic Force Requirements

We begin this chapter by examining the active-reserve mix in a non-stochastic environment in which the force requirements are the same in every period. For this analysis, we assume a force requirement of 20 units in each period. Therefore, we created a contingency in a new back-end named *Total20*, which has only a *Stabilization* state with a requirement of 20. We use this contingency in all runs described in this section.

For the runs described in this section, we use the forces defined earlier: *Active* and *ReserveIneff2*. In this section, we also choose to use the *Active4Reserve6* assignment rule defined in Chapter Five, which deploys reserve units 1.5 years in 6 (1 effective year) instead of 1 year in 6 (0.5 effective year). We copied this assignment rule from the back-end database used in Chapter Five.

The last step in setting up this analysis is to define a set of runs. To do this, we first selected a variety of different force structures to simulate. We designed each force structure so that it would be able to meet a requirement of 20 units on a 1-year-in-3 rotation for actives (2 years at home) and a 1.5-years-in-6 rotation (1 effective year) for reserves. Table 7.1 defines all of the force structures examined in this section. We began with the pre-transformation force of 43 active brigades and 34 reserve brigades. We then incrementally decreased the number of active brigades and added the number of reserve brigades necessary to continue to meet the requirement of 20 units (a trade-off of 2 reserve brigades for 1 active brigade). Table 7.1 shows the size of the active and reserve forces along with the percentage of the force that is active and the percentage growth in the reserves.

Table 7.1
Active-Reserve Force Structure Scenarios

Active Brigades/ Reserve Brigades	Percentage of Force That Is Active	Percentage of Growth in the Reserves
43/34	72	0
41/38	68	12
39/42	65	24
35/50	58	47
31/58	52	71
25/70	42	106

We defined a run for each of the six force structure scenarios defined in Table 7.1. After executing these runs, we compared the runs in terms of force stress (average time at home for actives) and cost (average cost per period). These results are shown in Table 7.2.

Table 7.2
Active-Reserve Mix Stress and Cost Results

Active Brigades/ Reserve Brigades	Average Time at Home for Actives (years)	Average Cost per Period
43/34	2.01	5,015
41/38	2.01	4,969
39/42	2.01	4,925
35/50	2.01	4,835
31/58	2.01	4,745
25/70	2.01	4,610

Table 7.2 shows that, for a steady force requirement, replacing active forces with reserve forces saves money in comparison to the current force (43 active, 34 reserve units). This is consistent with the results of Chapter Five, which show that, for the assignment rule that deploys reserve forces 1.5 years in 6, reserve forces are more cost-effective than actives in meeting a steady force requirement. However, as noted earlier, a steady force requirement is not very realistic and placing these forces in a stochastic environment can change the results significantly. This is explored in the next section.

7.2. Stochastic Force Requirements

In this section, we examine the same force structures defined in Section 7.1 but in a stochastic environment. This section shows that, in a stochastic environment, even though each force structure was defined to meet a constant stabilization requirement of 20, replacing active units with reserve units can increase stress on the active component.

We begin this analysis by creating a new stochastic contingency. We define the contingency to have three states: *Peace*, *War*, and *Stabilization*. We also define the force requirements in the *War* and *Stabilization* states to be 20 units, and the requirement in the *Peace* state to be zero. We use the same transition probabilities as used in Section 6.1 (shown in Figure 6.2). We then create six runs corresponding to the six force structure scenarios and set the initial condition for each run to *Peace*.

We execute each of the runs and examine the stress results (percentage of deployments with stressful reset times) and the cost results (average per-period cost), as shown in Table 7.3.

Table 7.3
Stochastic Stress and Cost Comparisons

Active Brigades/ Reserve Brigades	Percentage of Actives Home Less Than 1 Year Before Deployed	Percentage of Actives Home Less Than 1.5 Years Before Deployed	Percentage of Actives Home Less Than 2 Years Before Deployed	Average Cost per Period
43/34	0.00	1.67	5.00	4,487
41/38	0.00	1.81	5.64	4,380
39/42	0.21	2.18	8.29	4,276
35/50	1.09	2.19	18.03	4,072
31/58	2.99	7.06	19.74	3,867
25/70	8.11	11.49	25.68	3,557

Table 7.3 shows that, although replacing active units with reserve units decreases cost, it leads to an increase in stress on the active force. This occurs because, for stochastic contingencies, active units are needed to provide an immediate reaction force in response to an outbreak of war. If there are not enough fully rested active forces to provide this reaction force, this will lead to stress on the actives. Therefore, it is important to keep in mind that, although shifting forces from the actives to the reserves may save money, it can also lead to increasing stress on active units when faced with contingencies that arise without warning. It is likely that these results are sensitive to the duration and frequency of wars and stabilization operations. The SLAM program can be very useful in examining the effects of these parameters.

7.3. Summary

This chapter has shown that, under the given assignment rule and cost parameters, when force requirements are fixed, replacing active units with reserve units can save money while maintaining the same level of stress on the force. However, in a stochastic environment, although replacing active units with reserve units can save money, it can also increase the level of stress on the active force.

This chapter has shown how the RAND SLAM program can be used to help optimize the active-reserve mix. The analyses performed in this section are meant to be illustrative and show how a problem can be structured for analysis in the SLAM program. The results are also consistent with other results we have seen in this document. First, reserves are more cost-effective than actives for a steady force requirement (when they are deployed 1.5 years in 6). Second, active units are needed for a quick reaction. Therefore, shifting too many forces to the reserves places extra stress on active units to meet these requirements. In a more thorough analysis, one would examine these results under a variety of different threat scenarios, cost parameters, and assignment rules.

CHAPTER EIGHT

Conclusion

This report has demonstrated the power of the RAND SLAM program through a sequence of increasingly complicated simulation runs. Specifying, running, and interpreting a simulation run is nontrivial. The RAND SLAM program provides the required substantial software programming foundation.

The novel feature of the RAND SLAM program is its ability to simulate stochastic force requirements. This allows an analyst to explore how the varying nature of military requirements affects force structure decisions. Exploring this decision space can suggest force structures that are more or less attractive depending on the assumptions made by the analyst. As with any good simulation model, the RAND SLAM program forces analysts and policymakers to be explicit about their implicit understanding of the threat environment and their force allocation rules. The analyst must specify the following:

- How likely is each conflict?
- What forces are available?
- What is the relative cost of different types of forces?
- In what order should forces be utilized?

The model then allows the analyst to easily consider a wide range of alternative types of forces and force structures. Given the likelihood of conflict, the relative cost of different types of forces, and the level and type of forces, the model calculates

- cost
- risk (i.e., inability to provide required forces)
- stress on the forces.

A policymaker can then use this information to choose a combination of force Structures, force Levels, and force Allocation rules (the "S," "L," and "A" of "SLAM") that best balances cost, risk, and stress.

Of course, the preceding discussion is an idealized view of the policy process. Several caveats are worthy of note. First, the user of the SLAM program quickly becomes aware of his/her ignorance and the ignorance of the field. We simply do not know appropriate values for many of the parameters (e.g., the cost to recruit much larger or smaller forces). There is little consensus as to the likelihood of future conflicts (Knightian uncertainty). The relative effectiveness of reserve units is also a matter of debate. The RAND SLAM program allows the user to explore the parameter space to examine the effects of many of these uncertainties and also provides motivation for a new round of research to build a better foundation of program inputs.

Second, every model is incomplete. There are real-world constraints and costs that SLAM does not consider. Further versions of the model can incorporate such additional features. Some extensions should be relatively easy to incorporate, whereas others would require considerable reworking, and some are nearly computationally infeasible. We leave such extensions to a later version of the model.

A User's Guide to the SLAM Program

The body of this report has walked the user through several policy analyses using the SLAM program. Those analyses were intended to demonstrate the basic steps of performing an analysis and some of the features and options in the SLAM program. This appendix provides a more systematic and thorough introduction to the SLAM program.

A.1. Basic Sequence of Tasks

The basic sequence of tasks in running SLAM is as follows:

- *Install*: Install the program (once).
- *Begin*: Open the program and choose a data file.
- *Create*: As needed, create new conflicts, forces, and assignment rules (i.e., beyond those in an existing file or created by earlier user runs).
- *For each simulation run*
    - o *Specify*: Specify a SLAM run (i.e., which conflicts, which assignment rule, which level of forces, which training resources).
    - o *Run*: Execute the simulation run (i.e., instruct the program to compute the corresponding allocation of forces in each period).
- *Post-Process*: Create summary measures within and across simulation runs.
- *Write*: Write the results to a file (as needed).
- *Save*: Save a new simulation environment, including any new conflicts, forces, and assignment rules, but not including any specific run output (as needed).

The following sections of this appendix consider each of these steps as well as how to set global program options.

A.2. Install

The SLAM program is distributed as a CD[1] containing three Microsoft Access databases, an Excel spreadsheet, two GAMS scripts, and the MatrixVB installer. The three Microsoft Access databases include one front-end and two back-end databases. The front-end database

---

[1] To obtain the RAND SLAM distribution CD, contact RAND's Forces and Resources Policy Center using the information provided in the preface of this report.

for version 1 of the SLAM program is the file "Slam1.mdb." The SLAM distribution package comes with two general back-end databases: (1) a very basic set of data for use in this appendix (SLAM1be.mdb), and (2) a file that includes all of the contingences, forces, assignment rules, and run parameters used in the main part of this document (SLAM1Analysis.mdb). The SLAM distribution CD also includes an Excel spreadsheet called CreateAssignmentRule.xls for use in determining force allocation rules (see Appendix D for more details). The distribution CD also contains the files c3i.gms and c3l.gms, which include the GAMS program code. The file c3i.gms solves simulation runs as integer programs, whereas c3l.gms solves simulation runs as linear programs. The choice between these two must be made by the user when he or she defines a simulation run in Access. We suggest that the analyst not make any changes to the GAMS program files, as this could cause the SLAM program to fail. Lastly, the distribution CD includes the MatrixVB installer (matrixvb4520rt.exe), which installs the components necessary to perform the matrix calculations needed by the *Compute* button on the Run Parameters form (see Appendix B for more details).

We suggest that the analyst create a new folder called SLAM in the directory C:\Program Files and copy all files from the SLAM distribution CD to this folder. After this is done, the analyst is ready to begin using the SLAM program.

A.3. Begin

To begin using the SLAM program, the analyst should open the front-end database file Slam1.mdb. This brings up the SLAM opening screen. This screen provides the user with three options for opening a back-end database: *Default Data File*, *Last Data File*, and *Select Other Data File*, as shown in Figure A.1.

Figure A.1
SLAM Start-Up Screen

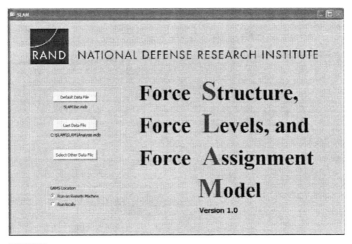

RAND *TR433-A.1*

The *Default Data File* is always set to the basic back-end in the distribution package: SLAM1be.mdb. The *Last Data File* is simply the last back-end database that was opened using the program. Finally, the *Select Other Data File* option allows the analyst to select any available file. The user must select a valid back-end database to continue.

The user may also choose where he wishes to run the SLAM program from two locations: (1) use a version of GAMS installed on his machine (run locally), or (2) use a GAMS version installed on a remote RAND machine that is hard-coded into the SLAM program and requires access to the RAND network (run on a remote machine). GAMS can be downloaded from http://www.gams.com/download/. A free version of GAMS is available, but is limited to 300 variables and 300 constraints. This version is not large enough to solve any of the problems in this report that involve reserve forces. A GAMS license with the features necessary to solve all of the problems contained in this report can be purchased at a cost of $11,200 commercially or $1,920 for an academic license (all prices as of November 27, 2006).

A.3.1. Create

Once a database has been opened from the SLAM opening screen, the user is presented with four options in the menu bar—File, Window, SLAM, and Help—as shown in Figure A.2.

Figure A.2
SLAM Toolbar

RAND *TR433-A.2*

The SLAM pull-down menu allows the user access to the data input forms. This section presents and discusses the key input screens that can be accessed from the SLAM menu.

A.3.2. Overview

The SLAM pull-down menu allows the analyst to set program parameters, edit input data, execute the program, and view/export results. As shown in Figure A.3, the menu items are: *Contingencies, Forces, Assignment Rules, Run Parameters, Run Batched, Reports and Analysis, Data Management, Options,* and *Exit.* Each of the first four options brings up a form that allows the analyst to edit the simulation parameters.

Figure A.3
SLAM Pull-Down Menu

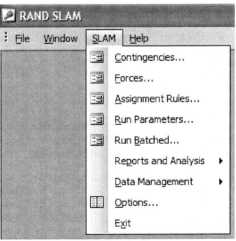

RAND *TR433-A.3*

Each of the forms to edit the simulation parameters includes a *New* and a *Copy* button. The *New* button creates a new item on the form with no retained information. The *Copy* button takes the currently selected item and creates a new entry with identical information. These two buttons provide the user flexibility when creating new simulation parameters.

It is important to note that time periods and force size are not explicitly defined in the SLAM program. For time periods, the user must choose units (e.g., months, quarters, years) and be consistent when entering all data. In the analyses contained in this report, we designated each period as a calendar quarter.

Similarly, for force size, the analyst must decide what level of detail he or she wishes to examine (e.g., brigade, unit, individual, etc.) and be consistent throughout. In the analyses of this report, we designated each unit as a brigade.

A.3.3. Contingencies

The Contingencies form allows the user to create and edit the types of contingencies that may arise (e.g., Iraq, Korea). For each contingency, the user must specify general characteristics, states and requirements, and a contingency transition matrix.

General Subform. The first tab on the Contingencies form shows the general characteristics of a given contingency. It allows the analyst to name and describe the contingency. This subform also contains a check box labeled "editable." On this and all other subforms, this box cannot be edited by the user and remains checked until a simulation is executed using the information on the specified subform. After a simulation is executed using this information, the editable check box is unchecked, and the user will be unable to make changes to the information for the selected parameters. This occurs because, after a run is executed, all of the parameters for that run are stored in the associated forms. These parameters are locked so that one can determine the parameters that were used for a specific run after the run was executed.

States Subform. The States subform allows the analyst to specify the number of units required for a given state of a specified contingency. This subform also allows the analyst to create new states. In the distribution package, there are three states: *Peace, War*, and *Stabilization*. Additional states can be added by clicking on the database window icon in the toolbar when any form is open and opening tblStateList. This table allows the user to enter new states and abbreviations (abbreviations are limited to five characters), which will then appear in the Contingencies . . . States pull-down menu. In the current version of the SLAM program, the user can enter up to a maximum of 24 states for each contingency.

For each of the states in each contingency, the number of units required for each type of requirement (Req0–Req5) must be specified by the user. The user must explicitly define what a unit is: A unit can be a brigade, company, person, etc., but the definition must be used consistently throughout the SLAM program.

There are six types of requirements hard-coded in the SLAM program, labeled Req0 through Req5. Each of these requirements adds a constraint that the total number of each type of effective force deployed must be greater than or equal to each requirement. The contribution of each force type to each requirement is specified on the Force Effectiveness subform. If there is only one type of force requirement, then all the analyst need do is fill in the requirements for Req0 and leave the zeroes in all the other requirements, as shown in Figure A.4.

Figure A.4
Contingencies: States Subform

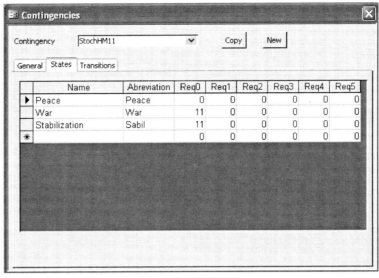

Contingencies: Transitions Subform. This subform allows the analyst to specify the transition probabilities from each of the states between any two time periods. This matrix of transition probabilities is the basis for the Markov process that determines the state of each of the contingencies in each time period. As displayed in Figure A.5, for a contingency with three states, the transition matrix is a 3×3 matrix, with each entry representing the probability of transitioning from the states on the left to the states on the top. For example, the entry in row 1, column 1 (*Peace, Peace*) is the probability that peace will continue in the current period given that the contingency was peaceful the last period.

Figure A.5
Contingencies: Transitions Subform

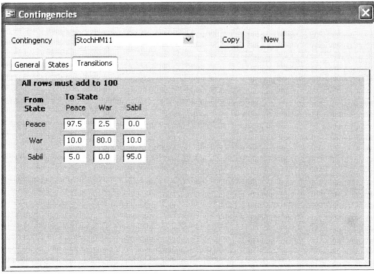

A.3.4. Forces

The Forces form allows the user to create and edit the types of forces available. Examples of force types are the active component and the reserve component. For each force type, the user must specify a name, description, cost, effectiveness by period of deployment, and training requirements by period of deployment.

General Subform. The first tab on the Forces form shows the general characteristics of a force. This tab is displayed in Figure A.6. The first two fields on this subform allow the analyst to name and describe the force. The next two fields on the subform require that the analyst specify the maximum number of periods that the force will be home or deployed. If the maximum number of periods home or deployed is less than the number of periods the simulation will run (specified in Run Parameters), the maximum period will be an absorbing state. This means that all units assigned to be at home or deployed longer than the maximum number of periods will be assigned to the maximum condition. As described in the body of this document, the appropriate choice of the maximum number of periods home and away will depend on the outcome of interest.

Figure A.6
Forces: General Subform

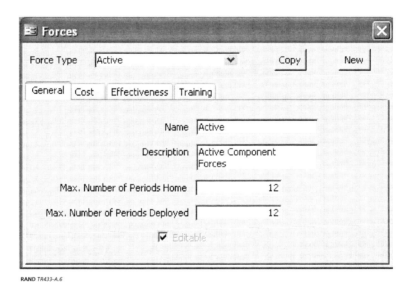

RAND *TR433-A.6*

Cost Subform. The Cost subform, shown in Figure A.7, allows the analyst to specify the cost of each force per period in each of three statuses: home, deployed, and training. Units of cost are not explicitly defined in the program, but the analyst should be consistent for all forces and types of cost.

Figure A.7
Forces: Cost Subform

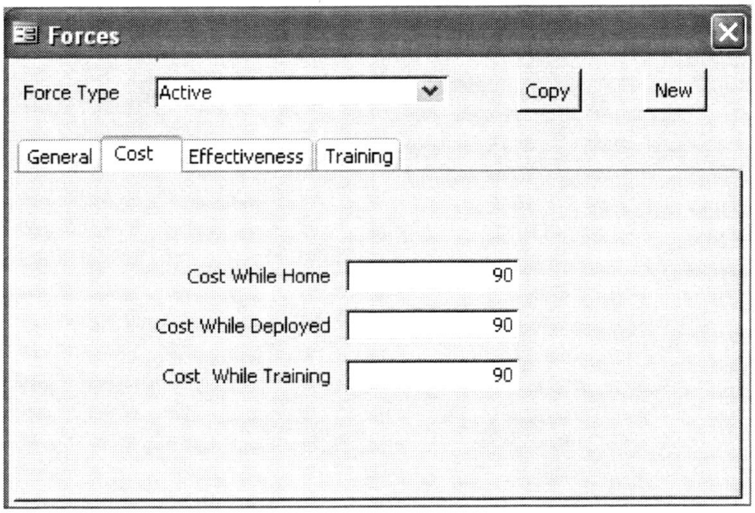

RAND *TR433-A.7*

Effectiveness Subform. In this subform, the analyst can specify the force effectiveness for each requirement type by period. Each requirement in the Contingencies: States subform is related to force effectiveness by a number. For instance, Req0 is associated with Eff0 and Req1 is associated with Eff1. The default value for force effectiveness is 0. This default value should be used for all requirements not used in the simulation. The analyst enters the start period (periods of deployment) in the first column, the end period in the second column, and the force effectiveness for each requirement type over the specified period in the third through eighth columns. Force effectiveness is measured on a 0 to 1 scale, with 1 being the most effective and 0 being completely ineffective. It is possible to scale any measure of force effectiveness from 0 to 1.

The six types of requirements and corresponding effectiveness allow the user to create one type of force and specify different types of capabilities within that force. For instance, the user can specify one type of force—for example, actives—and then specify different capabilities of that force to meet different types of requirements.

The effectiveness table is especially critical for forces such as reserves that are not effective in all periods of deployment. This is modeled by specifying that reserves contribute nothing to the requirement in train-up periods (the first two periods of deployment in the examples contained in this document) and then 1 in every period deployed thereafter, as shown in Figure A.8.

Figure A.8
Forces: Effectiveness Subform

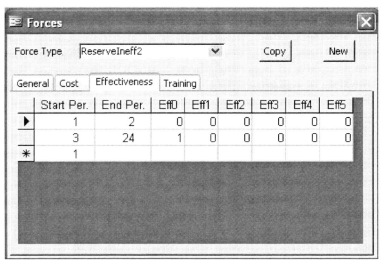

Force Training by Period. In this subform, the analyst can specify force training needs by period. The tables in these subforms have the same format as the Forces: Effectiveness subform, except that here the related constraints are for training. This subform allows the user to specify the periods of deployment in which a unit needs training of any of five types. For instance, Figure A.9 shows a reserve force that requires a full allotment (1 full unit) of type 1 training in periods 1 and 2 of deployment and no training in all other deployed periods. As with force effectiveness, training needs in each period are on a scale of 0 to 1.

Figure A.9
Forces: Training Subform

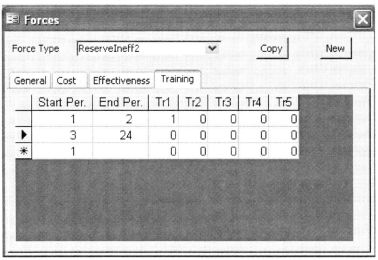

A.3.5. Assignment Rules

The Assignment Rules form provides the ability to specify different sets of assignment rules representing current or future DoD policies. For each set of assignment rules, the user must specify general characteristics, forces to use, and a force assignment rule.

General Subform. The General subform on the Assignment Rules form only requires the analyst to name and describe the rules he or she will define on the other subforms. No other specifications are needed.

Forces Subform. This subform, shown in Figure A.10, allows the analyst to choose which forces (active, reserve, etc.) will be used with this assignment rule. The analyst simply clicks on the desired forces to select them for use with this assignment rule.

Figure A.10
Assignment Rules: Forces Subform

RAND TR433-A.10

Assignment Rule Subform (Basic). This subform allows the analyst to specify the force assignment ordering that will be implemented during the simulation run. There are five columns in the table. The first column is rank, which defines the deployment ordering for the forces in ascending order (those with ranking 1 are deployed first). The second column is the force type, which is a drop-down menu of the forces selected on the Forces subform. The third column is the status of the force type (either home or deployed in this run). The fourth and fifth columns allow the analyst to group force types together by their current state (home one period, deployed one period, etc.). The status and periods refer to the condition of the force at the beginning of the period in which it is being considered for deployment. For instance, a simulation run using the assignment rule shown in Figure A.11 will first deploy those units previously deployed 3, 2, and 1 periods (in that order). The model will then deploy all units at home, starting with those at home the longest (in this case, 24 periods). Lastly, this model will deploy active units deployed 4 or more periods. This subform allows for easy grouping, but can only be used for forces that are fully effective in all periods. For a more detailed and flexible way to order the forces, the analyst must use the Advanced Assignment Rule subform.

Figure A.11
Assignment Rule (Basic) Subform

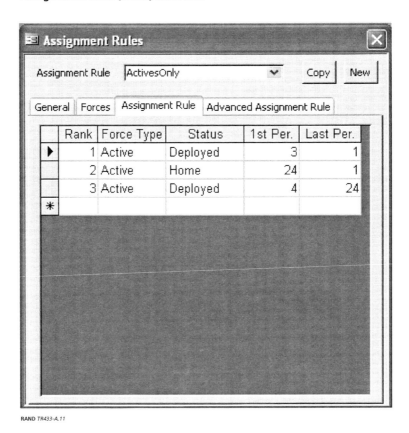

RAND *TR433-A.11*

Advanced Assignment Rule Subform. This subform allows the analyst to specify the exact force assignment ordering that will be implemented during the simulation run. There are four columns in the table on this subform. The first column gives the ordinal ranking of the different force conditions. The second column identifies the type of force. The third column identifies the status of the force type (home or deployed). The last column identifies how long the force has been in the current situation. This ordering means that the force type with the lowest ranking will be deployed first (if there are any units available), the second-lowest will be deployed second, and so on. The analyst can change the ordering in the list by using the up and down arrow buttons on the right. For example, the analyst might want active units that have been deployed three periods to be deployed again before any other units are deployed. To do this, the analyst would select "Active:Deployed:3" and then would click the up arrow until this row moves to the top of the list. This example is shown in Figure A.12. This assignment table is automatically updated by the Assignment Rule subform. Again, this ordinal ranking system only deploys units correctly for forces that are fully effective in all periods, such as the active component.

Figure A.12
Advanced Assignment Rule Subform

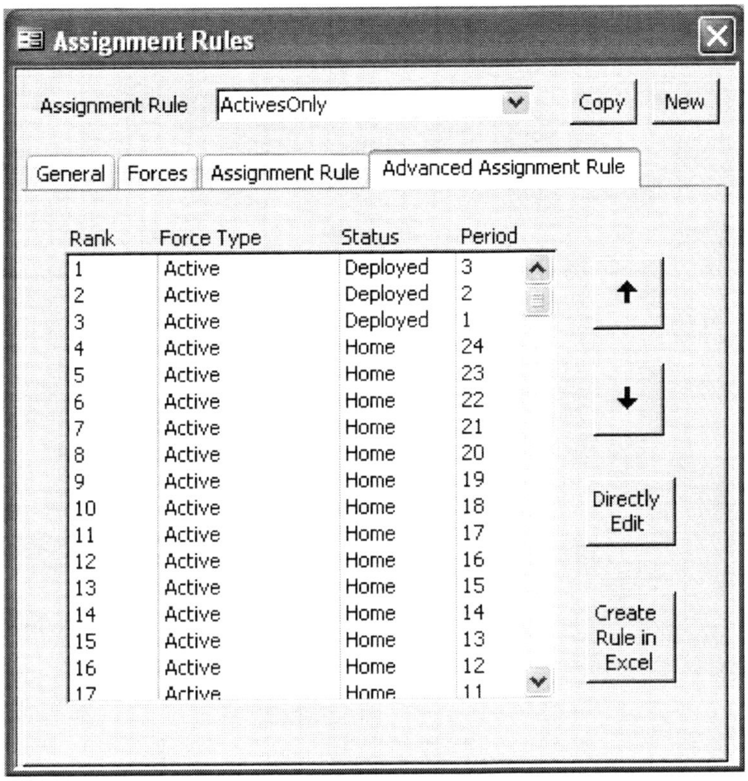

RAND TR433-A.12

When ranking forces that are not fully effective in all periods, more care must be taken to make sure that deployments occur as expected. The methodology for determining the rankings for these types of forces is described in Appendix D of this document.

The user has two options when specifying the assignment rule according to the methodology described in Appendix D. The first option, and the one that we recommend, is to use the built-in feature that opens an Excel spreadsheet with all of the tables needed to follow this algorithm. The user can use this feature by clicking on the *Create Rule in Excel* button on the Advanced Assignment Rule subform. This will open a spreadsheet with tables created based on the assignment rule parameters specified by the user. This spreadsheet is shown in Figure A.13; see Appendix D for more on how to use it. When the user is finished with the spreadsheet, he or she simply needs to close the spreadsheet and the results will be automatically imported into the back-end database that is currently open.

Figure A.13
Excel Assignment Rule Spreadsheet

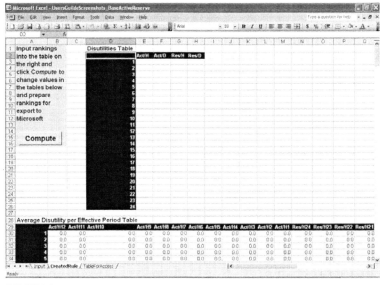

RAND *TR433-A.13*

The second option available to the user is to copy the assignment rule disutilities directly into the SLAM program by clicking on the *Directly Edit* button on the Advanced Assignment Rule subform. The force disutilities can be entered directly into the first column of the table titled "Order." Figure A.14 shows an example of a directly edited Deployment Disutilities form. Once each of the force-status combinations has been ranked, the user can click on the *Done* button, and he or she will see the newly entered force disutilities appear on the Advanced Assignment Rule subform.

Figure A.14
A Directly Edited Deployment Disutilities Form

DirecltyEditForceAssignment : Form

Directly Edit Force Assignment Order

	Order	Force	Home/Away	Period	
▶	3	Active	Deployed	1	
	2	Active	Deployed	2	
	1	Active	Deployed	3	
	28	Active	Deployed	4	
	29	Active	Deployed	5	
	30	Active	Deployed	6	
	31	Active	Deployed	7	
	32	Active	Deployed	8	
	33	Active	Deployed	9	
	34	Active	Deployed	10	
	35	Active	Deployed	11	
	36	Active	Deployed	12	
	37	Active	Deployed	13	

Done

RAND *TR433-A.14*

A.4. Specify

In the SLAM pull-down menu, the fourth option is titled *Run Parameters*. Clicking on this option opens a form that allows the analyst to choose the desired parameters for a simulation run and then execute that run. For each simulation run, the user must specify the general characteristics of the run, the contingencies to include, an assignment rule to use, force levels, initial conditions, and training constraints.

General Subform. This subform allows the analyst to specify the general characteristics of each simulation run. Figure A.15 shows an example Run Parameters: General subform. The first two fields on this subform allow the user to name and describe the run. The third field asks the user to specify the number of periods to execute the simulation. The fourth and fifth fields ask the user to specify the start and end periods for the run. The sixth and seventh fields allow the user to specify the parameters for the look-ahead periods, as described in Appendix C. The eighth field allows the user to specify the number of look-ahead periods (this is only necessary when using forces that are not fully effective in all periods). The ninth field allows the user to specify whether the simulation will solve as an integer or linear program. The final two fields allow the user to choose the type of look-ahead methodology to use: *Most-Likely* or *Tolerance*. If the user specifies tolerance, he or she must input a tolerance greater than 0 in the last field. Appendix C contains more information on the different look-ahead methodologies.

Figure A.15
Run Parameters: General Subform

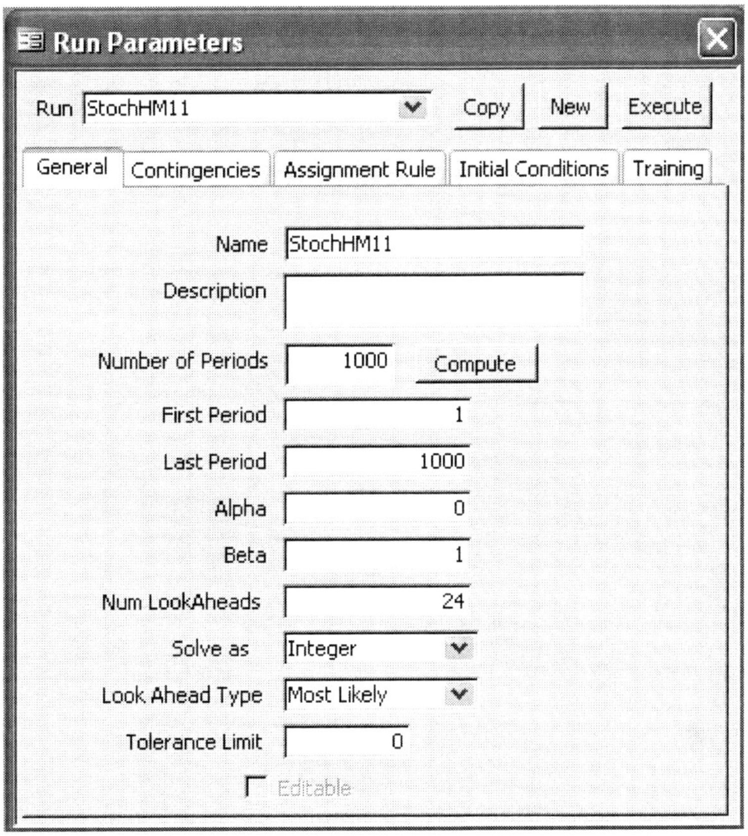

Contingencies Subform. This subform allows the analyst to select the contingencies to include in the simulation run. The analyst can select multiple contingencies to include. Contingencies are considered as independent events: If the analyst would like to include dependencies, he or she must define a new compound contingency that incorporates any dependency between contingencies. An example Run Parameters: Contingencies subform is shown in Figure A.16.

Figure A.16
Run Parameters: Contingencies Subform

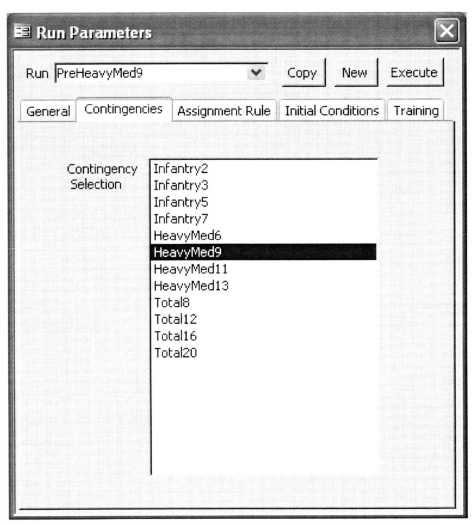

RAND *TR433-A.16*

Assignment Rule Subform. In this subform, shown in Figure A.17, the analyst can select the assignment rule to use for the specified simulation run. Once an assignment rule is selected, the analyst can change the number of units available for each force type included in that rule.

Figure A.17
Run Parameters: Assignment Rule Subform

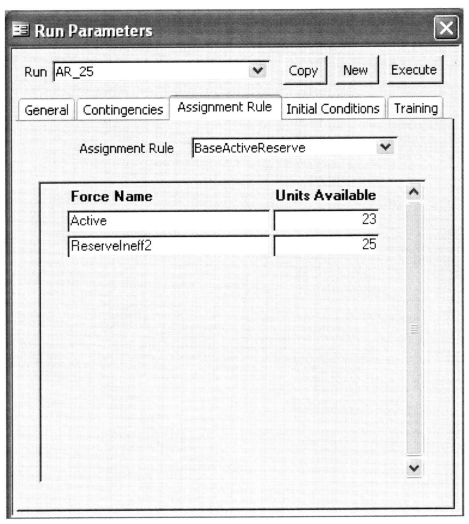

Initial Conditions Subform. This subform allows the analyst to choose the initial state for each of the contingencies selected on the Contingencies subform. The user can select each of the contingencies from the *Contingency* pull-down menu and then select the initial state using the *Initial State* pull-down menu. In Figure A.18, the contingency *StochHM11* would start period 1 in the *Peace* state.

Figure A.18
Run Parameters: Initial Conditions Subform

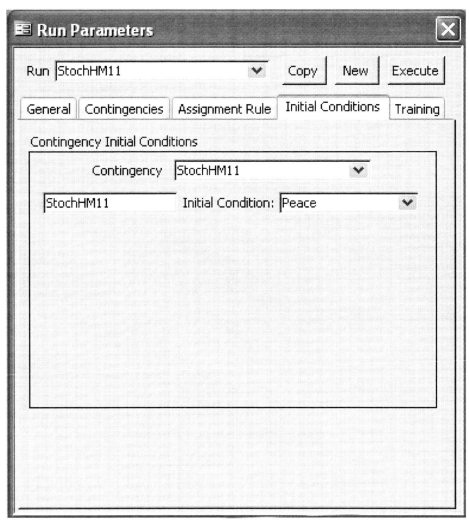

Training Subform. The Training subform allows the analyst to specify the training constraints. The training constraints are the maximum number of units that can be training in each type of training at any given time. For instance, in Figure A.19 there is a training capacity of 4, corresponding to *TR1:* This means that no more than 4 units (if each requires a full unit of training) can be trained at any given time at *TR1.* Training constraints are not used in any of the analyses contained in this document, but can provide a more realistic simulation if needed.

Figure A.19
Run Parameters: Training Subform

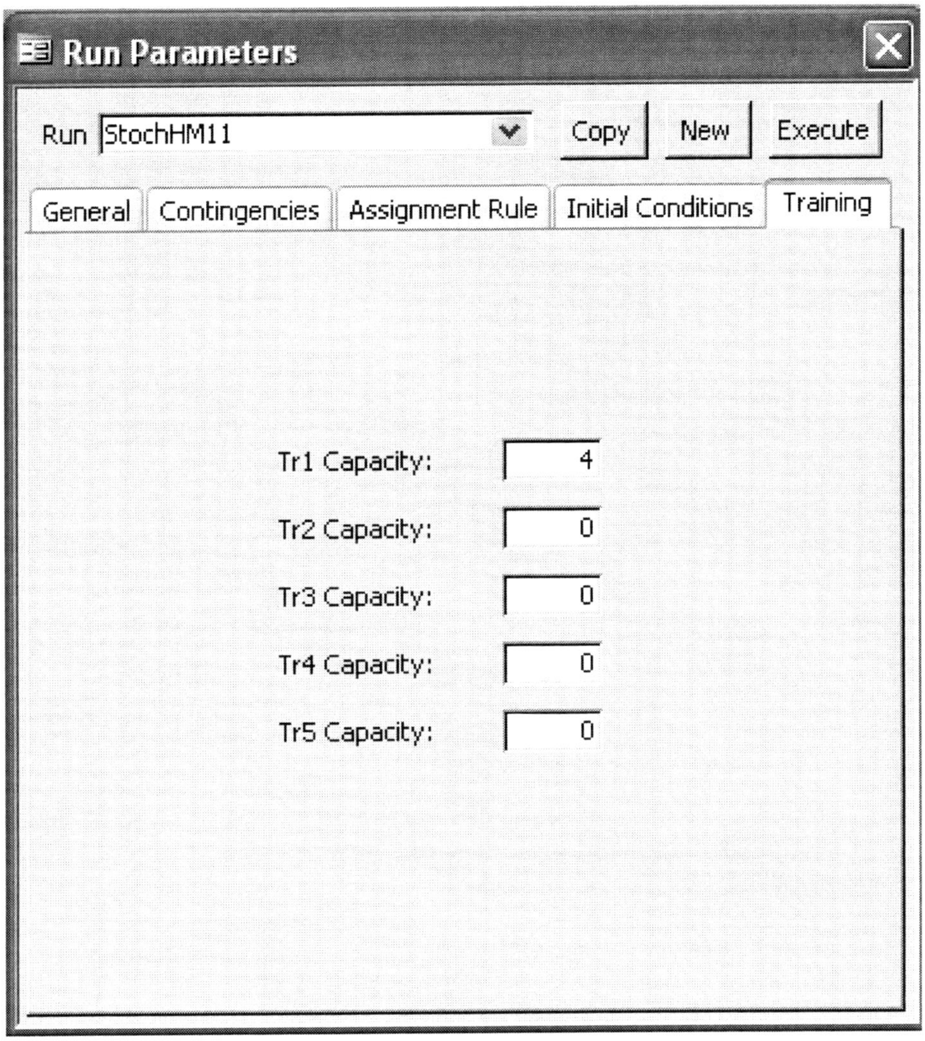

A.5. Run

The SLAM program has two options for executing runs. Each run can be executed individually from the Run Parameters form, or multiple runs can be executed from the Run Batched form.

Single Run. Once the analyst has specified the run parameters, the simulation is ready to run. A single simulation run can be executed from the Run Parameters form by selecting the run to be executed and then clicking on the *Execute* button. If the chosen run has not been executed before, the program will ask the analyst to confirm whether to execute this run, since all of the included parameters will no longer be editable. If the chosen run has been executed before, the program will return a message saying it has been executed before and asking for confirmation to run it again. Following affirmative responses to this prompt, the simulation will begin running and will display the message box shown in Figure A.20 when finished.

Figure A.20
Run Completed Message Box

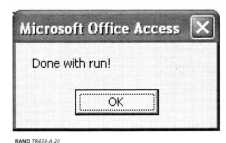

RAND *TR433-A.20*

Batched Runs. To allow the analyst to execute multiple runs concurrently, the SLAM program has the ability to execute runs in batches. This can be accomplished by opening the Run Batched form from the SLAM pull-down menu. This opens a table showing all of the existing runs in the database. Runs can be selected by setting the values in the first column to "yes" for the desired runs. The selected runs can then be executed by clicking on the *Submit Runs* button. Figure A.21 shows an example used in Chapter Three, in which only pre-transformation runs are executed.

Figure A.21
Run Batched Form

	Select Run?	Name	Periods	Look Aheads	Alpha	Beta
▶	Yes	PreInfantry02	1000	4	0	1
	Yes	PreInfantry03	1000	4	0	1
	Yes	PreInfantry05	1000	4	0	1
	Yes	PreInfantry07	1000	4	0	1
	Yes	PreHeavyMed06	1000	4	0	1
	Yes	PreHeavyMed09	1000	4	0	1
	Yes	PreHeavyMed11	1000	4	0	1
	Yes	PreHeavyMed13	1000	4	0	1
	Yes	PreTotal08	1000	4	0	1
	Yes	PreTotal12	1000	4	0	1
	Yes	PreTotal16	1000	4	0	1
	Yes	PreTotal20	1000	4	0	1
	No	PostInfantry02	1000	4	0	1
	No	PostInfantry03	1000	4	0	1
	No	PostInfantry05	1000	4	0	1
	No	PostInfantry07	1000	4	0	1
	No	PostHeavyMed06	1000	4	0	1
	No	PostHeavyMed09	1000	4	0	1

Submit Runs

RAND *TR433-A.21*

The SLAM program has different run times depending on where it is run. If the simulation is run on the local machine, it runs significantly faster than if it is run on a remote machine. As a rough approximation, in our experience the user should expect the simulation to work at a pace of about 30 minutes per 1,000 time periods. However, run times are dependent on the number of periods simulated and the number of variables in the linear program (determined by the number of look-ahead periods, number of forces, and maximum number of periods home/deployed for each force).

A.6. Post-Process

Once a simulation run has completed, the output data are available for viewing. There are four categories of run summary information: Contingencies, Costs, Force Assignment, and Force Stress. These can all be accessed from the SLAM menu in the Reports and Analysis submenu, as shown in Figure A.22. The model includes both Microsoft Access reports and Microsoft Data Access pages, which summarize the data and provide a format that can be easily exported into Microsoft Excel.

Figure A.22
Reports and Analysis Submenu

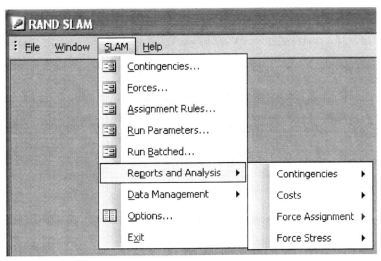

RAND TR433-A.22

Contingency Reports. The first three reports in the Contingencies submenu describe the contingency results. The first is a Microsoft Access report that shows how many times each contingency was in each state for each run. It also calculates the percentage of the total run time that each contingency is in each state. This can be seen in Figure A.23.

Figure A.23
Contingency Summary Report

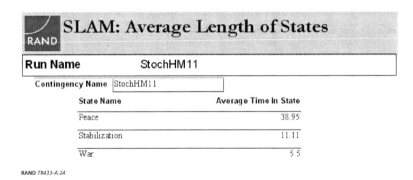

SLAM: Contingency Summary

Run Name	StochHM11

Contingency Name StochHM11

State Name	# Periods in State	% Periods in State
Peace	790	79.00%
War	110	11.00%
Stabilization	100	10.00%

Summary for Contingency Name = StochHM11 (3 detail records)

Sum	1000	100.00%

RAND TR433-A.23

The second contingency report is a Microsoft Access report that displays the average length of time spent in each state conditional on transitioning to that state. This report can be seen in Figure A.24.

Figure A.24
Average Length of States Report

SLAM: Average Length of States

Run Name	StochHM11

Contingency Name StochHM11

State Name	Average Time In State
Peace	38.95
Stabilization	11.11
War	5.5

RAND TR433-A.24

The third contingency report is a Microsoft Data Access page that contains the same data as the contingency summary report, but in a PivotTable that can be easily exported to Microsoft Excel, as shown in Figure A.25. This Data Access page also contains graphs that summarize the contingency data, as shown in Figure A.26.

Figure A.25
Contingency PivotTable

SLAM Contingency PivotTable			
Run Name ▼			
All			
	Contingency Name ▼		
	StochHM11		Grand Total
StateName ▼	# Periods in State ▼	% Time in State ▼	No Totals
Peace	790	79.00%	
Stabilization	100	10.00%	
War	110	11.00%	
Grand Total			

RAND *TR433-A.25*

Figure A.26
Contingency Graphs

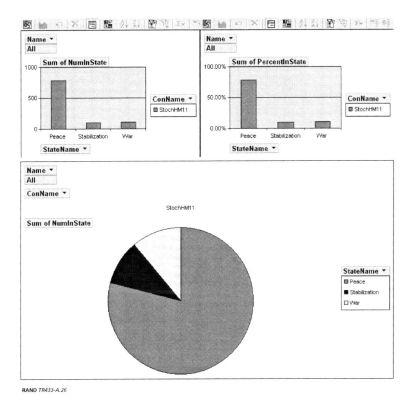

RAND *TR433-A.26*

The fourth contingency report includes a graph of the state of each contingency in each time period. This can be examined for a broad overview of how the state of each contingency changed over the course of the simulation run.

The final contingency reports show how force requirements change by period. The first report contains a PivotTable that displays the force requirements in each period, both in total and broken out by contingency. The second report contains two graphs that show how force requirements changed over time during the simulation run. The first graph shows the total number of units required by period. The second graph shows the number of forces required by each contingency in each period. These reports are not used in the analyses contained in this report but may be useful in future analyses.

Cost Reports. The second menu of reports provides the analyst with cost data for each simulation run. There are four different cost reports. The first cost report is the Cost Statistics report, which summarizes the key cost information for each run. This report provides only the average, minimum, and maximum costs for each run, as shown in Figure A.27.

Figure A.27
Cost Statistics Report

SLAM: Cost Statistics

RunName	StochFlex16		
	Avg. Cost Per Period	Min. Per Period Cost	Max. Per Period Cost
	4265.175	4200	5325

RunName	StochFlex20		
	Avg. Cost Per Period	Min. Per Period Cost	Max. Per Period Cost
	4287.15	4200	6600

RunName	StochHM11		
	Avg. Cost Per Period	Min. Per Period Cost	Max. Per Period Cost
	2265.75	2235	2985

RAND TR433-A.27

The second cost report is the Cost Summary report. It contains detailed cost information, showing the costs broken out by type of force and status for each run, as shown in Figure A.28.

Figure A.28
Cost Summary Report

SLAM: Cost Summary

Run Name	StochFlex16		
Force Name	Active		
	Cost at Home	Deployed Cost	Total Cost
Summary for Force= Active (1000 detail records)			
Sum	$3,298,950.00	$391,050.00	$3,690,000.00
Avg	$3,298.95	$391.05	$3,690.00
Min	$2,250.00	$180.00	$3,690.00
Max	$3,510.00	$1,440.00	$3,690.00
Force Name	ReserveIneff2		
	Cost at Home	Deployed Cost	Total Cost
Summary for Force= ReserveIneff2 (1000 detail records)			
Sum	$496,965.00	$78,210.00	$575,175.00
Avg	$496.97	$78.21	$575.18
Min	$285.00	$0.00	$510.00
Max	$510.00	$1,350.00	$1,635.00

RAND TR433-A.28

The third cost report shows even more detailed cost information, with costs broken down by period. The last cost report contains a graph of the average and maximum per-period cost by run.

Force Distribution. The Force Distribution report summarizes how each force is distributed in each period. The data is summarized in a PivotTable, which can be easily exported to Microsoft Excel. This can be used to verify that forces are being deployed as expected, as shown in Figure A.29. The columns in this table represent each of the periods for which the simulation was run. The rows in this table represent each of the possible states that a force can be in. For each force, there are rows for home and deployed, with the number of rows for each defined by the maximum number of periods entered by the user on the Forces form. The cell values in this table represent the number of units in each force-status-period in each period of the simulation.

Figure A.29
Force Distribution PivotTable

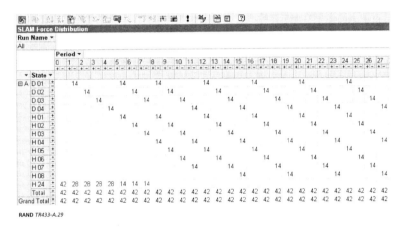

RAND TR433-A.29

Force Stress and Readiness Reports. The final category of reports displays measures of force stress and readiness. The first two force stress reports show the average amount of time units are home and deployed. Selecting *Report: Average Time Home/Deployed* from the Reports and Analysis . . . Force Stress submenu will open a Microsoft Access report similar to the one shown in Figure A.30.

Figure A.30
Average Time in State Before Transition Report

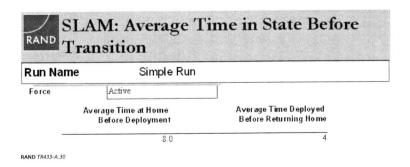

RAND TR433-A.30

The second report is a Microsoft Data Access page containing two bar graphs. The first graph plots the average time at home between deployments for all of the simulated runs in the back-end database. The second graph plots the average time deployed for all simulated runs.

The third report in the Force Stress submenu provides information on the number of ready units as defined by the user. This process is described in Section 3.7 of this document. Figure A.31 shows the report created after a ready unit has been defined by the user and a run has been executed.

Figure A.31
Summary Number of Ready Units Report

SLAM: Summary Number of Ready Units
RAND

Run Name		PostHeavyMed06			
	Force Name	Active			
		Number of Units in Each Readiness Level			
	Period	Readiness 1	Readiness 2	Readiness 3	Readiness 4
Summary for Force = Active (1000 detail records)					
Sum		10010	0	0	0
Avg		10.01	0	0	0
Min		4	0	0	0
Max		16	0	0	0
Summary for Run = PostHeavyMed06 (1000 detail records)					
Sum		10010	0	0	0
Avg		10.01	0	0	0
Min		4	0	0	0
Max		16	0	0	0

RAND TR433-A.31

The fourth and fifth reports in the Force Stress submenu display the percentage of units that experience stress. The fourth report shows the proportions of units that experience short reset times, as shown in Figure A.32. The fifth report shows the proportions of units that experience long deployments, as shown in Figure A.33. The levels of stress in both of these reports are hard-coded in the SLAM program, but they can be easily changed by anyone with a basic knowledge of Microsoft Access queries. The reset time reports are also defined only for actives and reserves. This means that, if the user wishes to use this report, the name of the active force must contain the word "active" and the name of the reserve force must contain the word "reserve." However, a user with some knowledge of Microsoft Access queries can create new stress definitions with little difficulty. The deployment length stress reports will show reports for all forces used in a run (not just active and reserves).

Figure A.32
Force Stress Report—Short Reset Time

SLAM: % Stressful Deployments: Reset Time

RAND

Run Name: StochFlex16 — * Time Periods in Years

Act. Home<1	Act. Home<1.5	Act. Home<2	Res. Home<4	Res. Home<4.5	Res. Home<5
0.00%	0.00%	0.04%	0.00%	0.00%	0.00%

Run Name: StochFlex20 — * Time Periods in Years

Act. Home<1	Act. Home<1.5	Act. Home<2	Res. Home<4	Res. Home<4.5	Res. Home<5
0.00%	8.32%	18.96%	0.00%	0.00%	0.00%

Run Name: StochHM11 — * Time Periods in Years

Act. Home<1	Act. Home<1.5	Act. Home<2	Res. Home<4	Res. Home<4.5	Res. Home<5
0.00%	13.62%	18.46%	0.00%	0.00%	0.00%

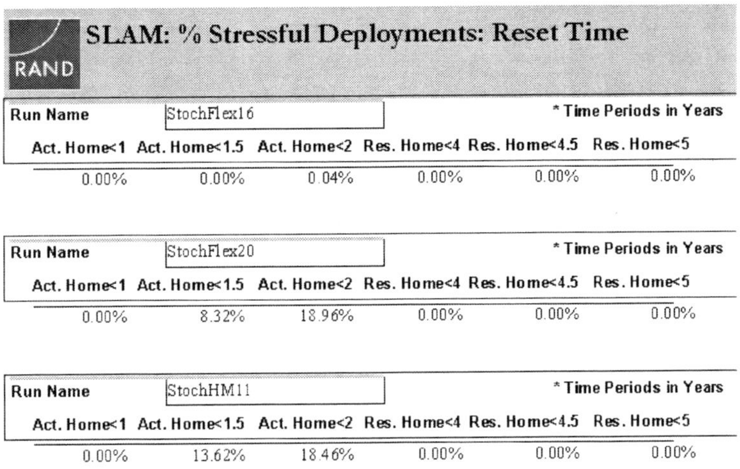

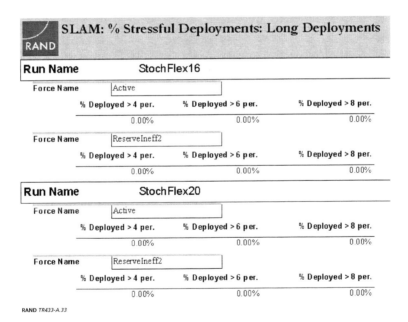

RAND *TR433-A.32*

Figure A.33
Force Stress Report—Long Deployments

SLAM: % Stressful Deployments: Long Deployments

RAND

Run Name StochFlex16

Force Name: Active

% Deployed > 4 per.	% Deployed > 6 per.	% Deployed > 8 per.
0.00%	0.00%	0.00%

Force Name: ReserveIneff2

% Deployed > 4 per.	% Deployed > 6 per.	% Deployed > 8 per.
0.00%	0.00%	0.00%

Run Name StochFlex20

Force Name: Active

% Deployed > 4 per.	% Deployed > 6 per.	% Deployed > 8 per.
0.00%	0.00%	0.00%

Force Name: ReserveIneff2

% Deployed > 4 per.	% Deployed > 6 per.	% Deployed > 8 per.
0.00%	0.00%	0.00%

RAND *TR433-A.33*

The final report in the Force Stress submenu shows the percentage of time that each force was home or deployed during each simulation run. This report is shown in Figure A.34.

Figure A.34
Force Stress Report—Percentage Time Home/Deployed

SLAM: % Time Home/Deployed		
Run Name	StochFlex16	
Force Name	Active	
	% Time Deployed	% Time Home
	10.60%	89.40%
Force Name	ReserveIneff2	
	% Time Deployed	% Time Home
	2.56%	97.44%
Run Name	StochFlex20	
Force Name	Active	
	% Time Deployed	% Time Home
	12.02%	87.98%
Force Name	ReserveIneff2	
	% Time Deployed	% Time Home
	3.42%	96.58%

RAND *TR433-A.34*

A.7. Write

To allow further analysis of simulation data, the analyst may wish to write the results to a file. This can be done from the Data Management submenu. Figure A.35 shows the three options available on this submenu.

Figure A.35
Data Management Submenu

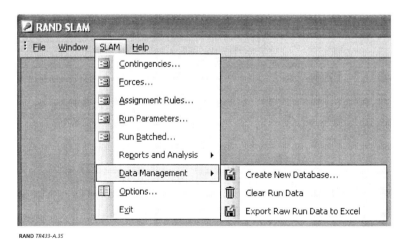

RAND *TR433-A.35*

The analyst can create Microsoft Excel files of the raw simulation output by selecting *Export Raw Run Data to Excel*. This will prompt the user for a file name and location for each of the three main raw output files. This data will appear in the specified Excel files in exactly the same format as it does in the Microsoft Access tables. However, we suggest that run data be kept within the SLAM program tables for further analysis.

The analyst can easily export any of the data in the PivotTables already described to Microsoft Excel using the *Export to Excel* button. However, since there is no cost PivotTable, there is an option in the Reports and Analysis . . . Cost submenu to export cost data directly to Excel. This will again prompt the user for a file name and location and will export an Excel file that includes all of the data used in the cost reports.

Lastly, the analyst can clear the back-end database of all previous simulation run data by clicking on *Clear Run Data* in the Data Management submenu. We recommend that the user either export all run data or save a copy of the current back-end database under a different name before clearing the run data from the back-end database.

A.8. Save

The analyst can save any part of the simulation environment by selecting the *Create New Database* option in the Data Management submenu, which brings up the screen shown in Figure A.36. This option creates a new back-end database. The user may select from any of the forces, contingencies, and assignment rules in the current back-end to be included in the new back-end database. After pushing the *Create* button, the user is prompted to enter a new file name and location, after which a new database is created.

Figure A.36
Create New Database Form

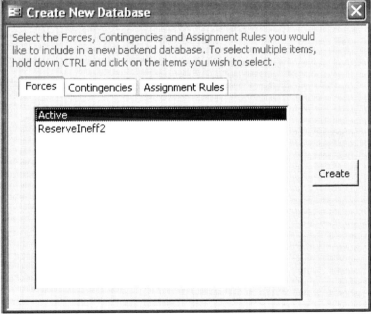

RAND *TR433-A.36*

A.9. SLAM Options

The second-to-last option available on the SLAM pull-down menu opens the SLAM Options form, as shown in Figure A.37. This form allows the user to specify global program options that are saved in the front-end database, rather than in the back-end database as with all other forms.

Figure A.37
SLAM Options Form

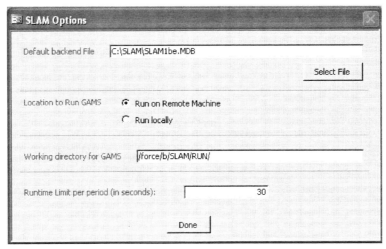

RAND *TR433-A.37*

This form includes four settings. Only advanced users should change these options. The first option defines the location of the default back-end database that is displayed on the SLAM start-up screen. In the distribution package, this will default to the back-end, SLAM1be.mdb. The second option allows the user to choose the location of the version of GAMS to be used (this option is also available on the SLAM start-up menu). The third option allows the user to select the working directory for GAMS. This is only relevant if the user chooses to run the simulation on a remote machine. The working directory defaults to a RAND UNIX directory. The user should be certain of the working directory before editing this option, because Microsoft Access does not have the capability of determining whether UNIX directories exist; entering an incorrect directory could result in an error during execution. The final option available to the user is the ability to specify a per-period run-time limit. The GAMS program will stop if it does not find a solution to a single period's problem in the specified amount of time. This can be useful in situations where multiple solutions exist and in situations where GAMS has trouble finding solutions. This option defaults to 30 seconds, which is more than adequate for all of the runs described in this document.

Technical Notes on the Implementation of the SLAM Program

This appendix provides a limited discussion of the technical issues in implementing the SLAM program. The discussion proceeds in three parts. First, we discuss the basic programming environment. Second, we discuss the process by which the SLAM program generates conflicts and describe how the program calculates the suggested number of periods based on the specified conflicts. Lastly, we briefly discuss the underlying linear program (LP) that computes the assignments. Appendix C contains further details as to the specification of the LP.

B.1. Programming Environment

The SLAM program is built using Visual Basic for Applications (VBA) for Access. The model consists of two parts: the VBA code and a primary Access database. In version 1 of the distribution package, the VBA code and core tables are in the file SLAM1.mdb, and the primary back-end Access database is in the file SLAM1be.mdb.

The Access database includes a large number of tables. At the conceptual level, there are tables for

- conflicts
- forces
- allocation rules
- run parameters
- per-period simulation output
- summary measures for each simulation run.

In practice, many of these tables have varying numbers of entries (e.g., the number of states for a conflict). They are therefore implemented as additional tables (see Appendix E for corresponding table names). The user should not need to know any of the specific table names. In fact, directly accessing the tables is discouraged, since doing so is likely to lead to corrupt data and program errors.

B.2. Conflicts

The RAND SLAM program assumes that contingencies arise according to a Markov process. This means that the probability of transitioning from peace to war in the current period does not depend on what state a contingency was in during the previous period (or in the period before that, etc.). This is a reasonable assumption for simulating most military threats. However, it is also possible that the probability of a second contingency transitioning from peace to war will be different depending on the state of the first contingency (being drawn into one war may increase the likelihood of a second war). It is possible to define contingencies in a way that can take into account these correlations, by specifying a compound conflict with multiple states (e.g., peace, war A, war B, war A and war B simultaneously).

The model can consider either short-term or long-term analysis. For short-term analyses, initial conditions matter. The smaller the transition probabilities and the larger the number of conflicts, the larger the required number of periods. As a rule of thumb, we choose the number of periods such that there will be approximately 100 periods in which all of the least-likely states from each conflict occur simultaneously (though sometimes this must be reduced due to run-time issues). As we noted in the body of this report, force-planning needs are often driven by those unlikely, but very high-requirement, states. For good analysis, it is crucial that the simulation "see" several such events.

The *Calculate* button on the Run Parameters: General subform allows the user to input the number of periods that he or she would like to "see" in the least-likely state (the default is 100). The SLAM program will use this information to calculate the number of periods for which the run should be simulated. To do this, the SLAM program first calculates the steady-state probabilities for each contingency. The steady-state probabilities can be determined by solving the linear equations below. In these equations, there are K contingencies indexed by k, so that the transition probability for a given contingency k from state i to state j can be written as p_{ij}^{k} and the steady-state probabilities for any state j of contingency k can be written as π_{j}^{k}. The number of states in each contingency is defined as S, so that $i=1,\ldots,S$ and $j=1,\ldots S$ for each contingency.

$$\pi_{j}^{k} = \sum_{i=1}^{S} \pi_{i}^{k} p_{ij}^{k} \qquad\qquad \sum_{i=1}^{S} \pi_{i}^{k} = 1$$

The first equation above defines $S-1$ independent linear equations. Utilizing the second equation, we can solve these S independent linear equations to find the S steady-state probabilities of any given contingency as described below.

The SLAM program uses matrix algebra to solve these linear equations. This requires a licensed and installed version of the MathWorks product MatrixVB (discussed in Appendix A). Define T as the transition matrix specified on the Contingencies form and P as a row vector of steady-state transition probabilities. We use the following 2 × 2 transition matrix as our example:

$$T = \begin{pmatrix} 0.8 & 0.2 \\ 0.5 & 0.5 \end{pmatrix}$$

If we name the states *Peace* and *War*, the steady-state row vector can be written (note the superscript is suppressed here because we are looking at only one contingency)

$$P = \begin{pmatrix} \pi_{p} & \pi_{w} \end{pmatrix}$$

In each of the following steps, we include the general matrix form of each equation on the left and the matrices corresponding to our example on the right. From the first linear equation above ($\pi_j^k = \sum\limits_{i=1}^{S} \pi_i^k p_{ij}^k$), we get that

$$PT = P \qquad\qquad (\pi_p \ \pi_w)\begin{pmatrix} 0.8 & 0.2 \\ 0.5 & 0.5 \end{pmatrix} = (\pi_p \ \pi_w)$$

With I defined as the identity matrix, this simplifies to

$$P(T - I) = 0 \qquad\qquad (\pi_p \ \pi_w)\begin{pmatrix} -0.2 & 0.2 \\ 0.5 & -0.5 \end{pmatrix} = (0 \ 0)$$

By definition, all rows of a transition matrix must add to 1. Therefore, there are only $S - 1$ linearly independent equations in the matrix $T - I$ shown above. We can reduce one of the columns to a column of zeroes as shown below:

$$(\pi_p \ \pi_w)\begin{pmatrix} -0.2 & 0 \\ 0.5 & 0 \end{pmatrix} = (0 \ 0)$$

The equation above has an infinite number of solutions. However, we also know that the steady-state probabilities must add to 1 (by the second linear equation: $\sum\limits_{i=1}^{S} \pi_i^k = 1$). In matrix form this means

$$P\begin{vmatrix} 1 \\ M \\ 1 \end{vmatrix} = \qquad\qquad (\pi_p \ \pi_w)\begin{pmatrix} 1 \\ 1 \end{pmatrix} = 1$$

We can augment the $T - I$ matrix by plugging in this new equation as shown below. The general form of this equation simply substitutes a column of 1s into the last column of the $T - I$ matrix and a 1 into the last column of the row vector on the right-hand side:

$$P(T - 1)_1 = (0 \ L \ 0 \ 1) \qquad\qquad (\pi_p \ \pi_w)\begin{pmatrix} -0.2 & 1 \\ 0.5 & 1 \end{pmatrix} = (0 \ 1)$$

The last step requires taking the inverse of the augmented $T - I$ matrix to solve for the steady-state probabilities. The SLAM program uses the MatrixVB add-in included in the distribution package to perform this calculation:

$$P = (0 \ L \ 0 \ 1)[(T - I)_1]^{-1} \qquad\qquad (\pi_p \ \pi_w) = (0 \ 1)\begin{pmatrix} -0.2 & 1 \\ 0.5 & 1 \end{pmatrix}^{-1}$$

After performing the multiplication on the right-hand side of the equation above, the result is a 1×2 row vector containing the steady-state probabilities. For our example, it turns out as follows:

$$(\pi_p \ \pi_w) = (0 \ 1)\begin{pmatrix} -1.4 & 1.4 \\ 0.7143 & 0.2857 \end{pmatrix} = (0.7143 \ 0.2857)$$

The methodology described above will work for all irreducible transition matrices, that is, the matrix cannot have any noncommunicating states. Given the types of analyses the SLAM program is designed for, this should never be a problem. This methodology is repeated for each contingency included in each simulation run.

Once the steady-state probabilities have been determined, the SLAM program determines the expected recurrence time for the situation in which all of the least-likely states from each contingency occur at once. This is necessary because this least-likely state will be the driving force in determining how many periods a simulation run must be executed for. The expected recurrence time is calculated in the equation shown below, using the steady-state probabilities calculated above. This equation takes the smallest steady-state probabilities from each of the contingencies and multiplies them together. Taking the inverse of this quantity provides the expected amount of time it will take for this rarest state to occur and recur thereafter (Hillier and Lieberman, 2005):

$$\text{Expected Recurrence Time} = \frac{1}{\prod\limits_{k=1}^{K} \min\limits_{j}\{\pi_j^k\}}$$

After clicking the *Compute* button, the user specifies the number of periods that he or she would like to "see" in the least-likely state. In the equation below, we define this as L. The SLAM program takes this value and multiplies it by the expected recurrence time to determine the number of periods to execute the run:

$$\text{Number of Periods} = \frac{L}{\prod\limits_{k=1}^{K} \min\limits_{j}\{\pi_j^k\}}$$

B.3. Solving the Linear Program

The core of the model is a linear program. Initially, we solved the problem using Excel's built-in "Solver" from Front Line Systems. A free version of Solver is limited to 200 variables. Our initial one-force LP using Solver had only 12 variables. A second LP using Solver grew to 48 variables with the addition of another force and a look-ahead period.

More complicated simulations (e.g., many look-ahead periods) require much, much larger models. It is easy to generate policy-relevant simulations with tens or even hundreds of thousands of variables. This raises a software issue and a hardware issue.

For LP software, we moved from Solver to the General Algebraic Modeling System (GAMS—http://www.gams.com/). GAMS allows not only a greater number of variables but also more flexibility in problem specification.

Even with appropriate software, solving LPs of this size requires considerable computational resources. They will solve on a standard PC, but they require a very large amount of time (30 minutes for a 1,000-period simulation). They will solve faster on a faster PC (e.g., one with more memory, faster processor) and even faster on a dedicated workstation.

Specification of the Integer and Linear Programs

The core of the SLAM program is an integer/linear program that implements the force allocation rule. Anyone considering using or extending the SLAM program needs to understand, at least at a heuristic level, what SLAM does.

In principle, the force allocation problem has an exact solution. However, that exact solution appears to be computationally infeasible. An analogy to chess is useful. In principle, there is an optimal set of chess moves that will never lose. In practice, there are so many possible moves that it is not possible to compute that optimal rule. Instead, chess programs choose moves by heuristics. This is also what SLAM does.

C.1. N-Step Look Ahead and Delayed Utility

Specifically, SLAM solves a sequence of N-step look-ahead problems. In a single period, SLAM optimizes the allocation of forces over N look-ahead periods, which produces an optimal allocation of forces in the current period given the "likely" requirements over the next N periods. This appendix includes a description of how the SLAM program determines "likely" requirements. This process is repeated in each period of the simulation. This methodology allows the SLAM model to deploy forces that would not generate any effective forces in this period but that would generate forces *that would be used* in some future period. If the user is including forces that are not fully effective in all periods of deployment, he or she must specify a number of look-ahead periods. For the analyses performed in this document with reserve forces, the number of look-ahead periods is set equal to the cycle length of the reserves (24 periods). This was found to provide reliable solutions, though other specifications may produce the same results.

This approach usually yields the desired result, but sometimes it does not. In those cases, it is often useful to specify some other constraint. Please report anomalous results for further consideration by the SLAM team.

C.2. The Resulting IP/LP Problem

The optimization problem is specified as follows. There is a variable for every combination of initial state (currently home or deployed for X number of periods) and possible transition (home or deployed in the next period) for every look-ahead period.

For example, the simulations presented in this report have two types of forces—actives and reserves—with cycle times of 12 and 24 quarters, respectively. Each force can make two transitions (home or deployed next period) from an initial state (currently home or deployed X periods). With reserves on a 24-quarter cycle, we have found that a 24-step look-ahead is sufficient. Thus, this simulation run has 3,456 variables (2 forces × 2 transitions × (12 + 24) initial states × 24 look-ahead periods).

There are constraints for the required number of units from each force, the possible transitions for each variable, and the number of units that can be training. Finally, there is a nonnegativity and a user-defined option to impose an integer constraint on each variable. The integer constraint is used in the analyses in this report because the concept of unit cohesion implies that units be sent over as a full team and not as fractions of a team. This constraint does not significantly increase run times for the analyses performed in this report, but can be easily relaxed by changing the *Solve as* option on the Run Parameters: General subform.

Running times are on the order of 2 seconds per period (with 24 look-ahead periods) and 30 minutes for 1,000 periods.

C.3. Formalizing the LP

In order to articulate the LP problem formally, we consider the most general situation by using the representations below. Each of these calculations is performed for each time period.

Parameters

There are a number of parameters that are specified for each simulation run. The first set of constants determines the weighting of the "disutilities" of forces in all look-ahead periods. When creating an assignment rule, the user creates "disutilities" of deployment for forces in the *first* look-ahead period (see Appendix D for more detail). The SLAM program calculates the disutilities in all future look-ahead periods by linearly weighting the disutilities from the first look-ahead period. α and β are the additive and multiplicative constants used in this linear weighting calculation:

α: additive constant used in calculating utilities for all look-ahead periods

β: multiplicative constant used in calculating utilities for look-ahead periods.

In all of the analyses in this report, we found that setting α equal to zero and β equal to one provided reliable solutions. Other specifications may also produce reliable results.

The second grouping of constants shown below determines the greatest number of periods that a force can be in a given state (home/deployed) for actives and reserves. In general, there would be a constant for every force and every state: P_{fc}. The user inputs these parameters on the Forces: General subform. The largest value is an absorbing state. For just active and reserve forces, the constants are

P_{ah}: the maximum number of periods actives can be at home

P_{rh}: the maximum number of periods reserves can be at home

P_{ad}: the maximum number of periods actives can be deployed

P_{rd}: the maximum number of periods reserves can be deployed.

Indices

f: force type (active/reserve)

c: status in current period (home/deployed)

n: status in next period

p: number of periods in current status (1, 2, etc.)

l: look-ahead period (1, 2, etc.). *l*=1 is the current period after which *l*=*n* is the (*n*–1)th look-ahead period.

i: force contribution/requirement type (hard-coded as i=0...5: req0...req5, eff0...eff5)

Functions

X[f,c,n,p,l]: These are the number of forces with characteristics *f, c, n,* and *p* in look-ahead period *l*. These are also the LP variables that are varied to find the optimal solution (the number of each force type in each state: for example, home one period and deployed the next in look-ahead period *l*).

U[f,c,n,p,l]: These are the disutilities. For *n* = deployed, these are the disutilities from the force assignment rule. For *n* = home, these are all equal to zero. The Force Ranking table determines the disutility of forces in the current period (*l*=1). For all subsequent look-ahead periods, the disutilities are calculated as follows:

$$\forall l > 1 : U(f,c,n,p,l+1) = \alpha + \beta U(f,c,n,p,l)$$

$E_i[f,c,n,p,l]$: The effectiveness of forces (how much each force type contribution *i* in each state contributes to each requirement *i*). For *n* = home, the effectiveness for all forces is equal to zero. For *n* = deployed, the effectiveness for the current period (*l*=1) is determined from the Forces: Effectiveness subform. For all subsequent look-ahead periods, the effectiveness is equated to the corresponding state from the previous look-ahead period:

$$\forall i, l > 1 : E_i(f,c,n,p,l+1) = E_i(f,c,n,p,l)$$

In the current version of the model, it is assumed that force effectiveness does not change over the course of a simulation run. This assumption could be relaxed with some significant modifications to the program.

$R_i[l]$: The force requirement of type *i* in look-ahead period *l*. The requirements in the look-ahead periods can be calculated in two different ways, either using the *Most Likely* method or the *Tolerance* method. The user can select which method to use on the Run Parameters form by changing the *Look Ahead Type*. The process of determining requirements in the look-ahead periods is described below for both the *Most Likely* and *Tolerance* methodologies.

The *Most Likely* look-ahead methodology determines force requirements in the look-ahead periods, as follows:

1. A random draw determines the state of each contingency in the first look-ahead period.
2. The corresponding requirements for the state of each contingency are summed to give the total requirements for the first look-ahead period (for each force contribution).
3. The requirements for each contingency in the second and all following look-ahead periods are determined by performing the following steps:
   a. Determine the state of each contingency in the previous look-ahead period (first period determined from random draw).
   b. Raise the transition matrix for each contingency to the $l - 1$ power, where l is the current look-ahead period. For instance, for the third look-ahead period, raise the matrix to the second power.
   c. For each multiplied contingency transition matrix, look across the row corresponding to the state of that contingency in the previous look-ahead period, find the state with the largest probability (if there is a tie, choose the first one), and choose that state as the state in the current look-ahead period.
   d. Identify the requirements for each contribution in the chosen state of each contingency and sum them to determine the requirements for each contribution in that look-ahead period.
   e. Repeat for all look-ahead periods.

The second look-ahead methodology available to the user is the *Tolerance* methodology, which determines force requirements based on a user-input *Tolerance Limit* on the Run Parameters form. The tolerance limit is the percentage of likely requirements in each look-ahead period that the user would like to be able to meet. The extreme case is 100 percent, which implies that the requirement in the next look-ahead period should be set to the maximum possible requirement for all feasible states of the world in the next look-ahead period.

In cases where the program cannot meet force requirements (there are not enough trained-up forces available), the program will return −1 for all decision variables in that period. The user may wish to avoid this problem by creating "phantom" forces that can fill requirements that cannot be met. Analyzing the usage of these "phantom" forces allows the user to calculate the ability of the forces to meet military requirements.

The *Tolerance* look-ahead methodology determines force requirements in the look-ahead periods, as follows:

1. A random draw determines the state of each contingency in the first look-ahead period.
2. The corresponding requirements for the state of each contingency are summed to give the total requirements for the first look-ahead period (for each force contribution).
3. The requirements in the second and all following look-ahead periods are determined by performing the following steps:
   a. Raise the transition matrix to the $l - 1$ power (where l is the current look-ahead period).

b. Find the probabilities of all states of the world in the current look-ahead period based upon the calculated transition matrices and the state of the world in the first look-ahead (always use the state of the first look-ahead period for all subsequent look-ahead calculations).

c. Rank the possible states of the world from most to least likely. Impose a cutoff where the cumulative likelihood of the most likely events is greater than or equal to the tolerance (specified by the user).

d. From among the most-likely events, choose the one with the largest requirement for each force contribution and use these as the requirements for the current look-ahead period.

e. Repeat for all look-ahead periods.

$S[f,c,p,l]$: The stock of forces in a given state. The stock variable helps us formalize the LP constraint that every unit in this period must be somewhere in the next period. For the current period ($l=1$), the stock variable is determined by the distribution of forces from the previous period. For the first period, this variable is determined by the initial distribution of forces defined for the simulation run. Currently, the model begins with all forces at home for the longest number of periods (p_{fh}). For all subsequent look-ahead periods ($l>1$), the stock variable is defined differently for different values of p (the number of periods in the current state [c]), as shown by the definitions below:

1. The number of units currently at home for one period must be equal to the sum of the units who were deployed last period and came home:

$$\forall f, \forall l > 1 : S\big[f,h,1,l\big] = \sum_{p=1}^{P_f} X\big[f,d,h,p,l-1\big]$$

2. The total number of units currently deployed for one period must be equal to the sum of those who were home last period, then deployed:

$$\forall f, \forall l > 1 : S\big[f,d,1,l\big] = \sum_{p=1}^{P_f} X\big[f,h,d,p,l-1\big]$$

3. The total number of units currently home for more than one period must be equal to the number of units home yesterday for one fewer period who did not deploy (who were assigned to be home again):

$$\forall f, \forall 1 < p < p_{fh}, \forall l > 1 : S\big[f,h,p,l\big] = X\big[f,h,h,p-1,l-1\big]$$

4. The total number of units currently deployed for more than one period must be equal to the number of units deployed yesterday for one fewer period who did not come home (who were assigned to be deployed again):

$$\forall f, \forall\, 1 < p < p_{fd}, \forall l > 1 : S\left[f,d,p,l\right] = X\left[f,d,d,p-1,l-1\right]$$

5. The number of units currently at home the longest (p_{fh}) is equal to the number of units home for one fewer period last period who are assigned to be home again, plus the number of units home for the same number of periods last period who are assigned to be home again. This state occurs because p_{fh} is an absorbing state in which units that are already at home this long and are allocated to remain home again stay in this state:

$$\forall f, \forall l > 1 : S(f,h,p_{fh},l) = X(f,h,h,p_{fh}-1,l-1) + X(f,h,h,p_{fh},l-1)$$

6. The number of units currently deployed the longest (p_{fd}) is equal to the number of units deployed for one fewer period last period who are assigned to deploy again, plus the number of units deployed the same number of periods last period who are assigned to deploy again. Again, this occurs because p_{fd} is an absorbing state:

$$\forall f, \forall l > 1 : S(f,h,p_{fd},l) = X(f,d,d,p_{fd}-1,l-1) + X(f,d,d,p_{fd},l-1)$$

Using the notation and definitions above, we can now formalize the LP. For each part of the LP below, the most general form of each equation is followed by the form relating to the problem addressed in this report with only actives and reserves.

Objective Function

The objective function is the dot product of the decision variables and their disutilities. More specifically, the objective function multiplies the disutility value of each state by the number of forces in that state and then adds up the products. The integer/linear program then attempts to minimize the objective function by changing the number of forces in each state subject to the constraints described below:

$$\min{}_X U[f,c,n,p,l] \bullet X[f,c,n,p,l]$$

For two force types (a=actives, r=reserves), the objective function is

$$\min{}_X \sum_{l=1}^{L} \left\{ \begin{array}{l} \displaystyle\sum_{p=1}^{P_{ah}} U\left[a,h,h,p,l\right] X\left[a,h,h,p,l\right] + \sum_{p=1}^{P_{ah}} U\left[a,h,d,p,l\right] X\left[a,h,d,p,l\right] \\[2em] +\displaystyle\sum_{p=1}^{P_{ad}} U\left[a,d,h,p,l\right] X\left[a,d,h,p,l\right] + \sum_{p=1}^{P_{ad}} U\left[a,d,d,p,l\right] X\left[a,d,d,p,l\right] \\[2em] +\displaystyle\sum_{p=1}^{P_{rh}} U\left[r,h,h,p,l\right] X\left[r,h,h,p,l\right] + \sum_{p=1}^{P_{rh}} U\left[r,h,d,p,l\right] X\left[r,h,d,p,l\right] \\[2em] +\displaystyle\sum_{p=1}^{P_{rd}} U\left[r,d,h,p,l\right] X\left[r,d,h,p,l\right] + \sum_{p=1}^{P_{rd}} U\left[r,d,d,p,l\right] X\left[r,d,d,p,l\right] \end{array} \right\}$$

Constraints

The IP/LP has three types of constraints it must satisfy while minimizing the objective function. First, the total number of each type of deployed forces must satisfy the military requirement. Second, the stock variable is used to make sure that every unit defined as being at home or deployed during this period must be somewhere the next period. Third, all the decisions variables (X) are constrained to be non-negative. The user also has the option of imposing an integer constraint on all decision variables. Each of these constraints is formalized below.

The first constraint specifies that the deployed forces must satisfy the military requirement. This constraint is formalized by saying that the dot product of the decision variables with the effectiveness must be greater than or equal to the force requirement for each type of requirement and force contribution (i) in each look-ahead period (l):

$$\forall l,i: R_i[l] \le E_i[f,c,n,p,l] \bullet X[f,c,n,p,l]$$

For only active and reserve forces, this expands to the equation below (variables for which n = home are not included because they provide zero effectiveness):

$$\forall l,i: R_i[l] \le \begin{cases} E_i[a,h,d,p,l] \bullet X[a,h,d,p,l] + E_i[a,d,d,p] \bullet X[a,d,d,p,l] \\ + E_i[r,h,d,p] \bullet X[r,h,d,p,l] + E_i[r,d,d,p] \bullet X[r,d,d,p,l] \end{cases}$$

The force requirements in the first look-ahead period are determined by the Monte Carlo process and user-defined requirements. The force requirements in all other look-ahead periods ($l>1$) are determined based on the look-ahead methodology chosen by the user on the Run Parameters form (*Most Likely* or *Tolerance*; see earlier discussion for details).

The second constraint is that the stock of forces at home for a given amount of time (p periods) in the current period is equal to the sum of units home for p periods that remain home next period and those who are deployed next period. For example, this means that the stock of forces at home for one year in the current period is equal to the sum of forces home for one year who will be deployed next period and those who will remain home next period:

$$\forall f,p,l: S(f,h,p,l) = X(f,h,h,p,l) + X(f,h,d,p,l)$$

Similarly, for units deployed in the current period, the stock of those deployed for some length of time is equal to the sum of those who will remain deployed next period and those who will come home next period:

$$\forall f,p,l: S(f,d,p,l) = X(f,d,d,p,l) + X(f,d,h,p,l)$$

The last constraint is that all the decision variables must be non-negative integers. The non-negativity constraint is built into the program, but the integer constraint is set by the user on the Run Parameters form and can be relaxed easily.

$$\forall f, \forall c, \forall n, \forall p, \forall l: X[f,c,n,p,l] \ge 0 \text{ and integer}$$

As discussed earlier, the integer constraint did not significantly increase run time for the analyses performed in the body of this report. However, in general, integer programs can take longer to solve than the corresponding linear program. If run time becomes an issue, a simulation can be run as a linear program (dropping the integer constraint) simply by selecting the *Linear* option on the Run Parameters: General subform.

C.4. Implementing the IP/LP

The constrained minimization problem formalized above has been implemented in GAMS and can be solved as either an integer or linear program. The GAMS code has two parts: an input template and a core program, each of which is discussed separately.

GAMS Input Template

The input template contains all of the necessary information to solve the IP/LP. It is created by the SLAM program for each run, in a file named "c3.gms." It contains the inputs in the order shown below. Following each input below is sample code for a simple example with two types of forces, two states, 24 look-ahead periods, 24 periods within each state, and one contingency.

1. *Per-Period Run-Time Limit (seconds)*:
   ```
   option reslim=30;
   ```
2. *Look-Ahead Methodology and Tolerance*:
   ```
   Scalar tolerance / 0 /
   ```
   This parameter is set to 0 if the most-likely methodology is to be used; any number other than 0 signals the use of the *Tolerance* methodology and specifies the tolerance limit.
3. *Force Type*:
   ```
   Set f /Active, Reserve/;
   ```
4. *Current Force Status*:
   ```
   Set c /Home, Deployed/;
   ```
5. *Number of Periods Home or Deployed*:
   ```
   Set p /p01*p24/;
   ```
6. *Number of Look-Ahead Periods*:
   ```
   Set l /L1*L24/;
   ```
7. *Force Contribution Dimensions*:
   ```
   Set fc / fc0 * fc5 /;
   ```
8. *Training Base Types*:
   ```
   Set t / tr1 * tr5 /;
   ```
9. *Disutility*:
   ```
   Parameter Util(f,c,n,p,l)
   /Active.Home.Deployed.p1.L1=999999
    Active.Home.Deployed.p2.L1=8000000

    ...

    Reserve.Deployed.Deployed.p24.L1=500/;
   ```

10. *Additive and Multiplicative Constants for Look-Ahead Utility:* These make sure that units that have a deployment lag time are considered (i.e., reserve units that are not effective until the third period of deployment):

```
Scalar alpha /0/;
Scalar beta /1/;
```

11. *Number of Units Initially in Each State:*

```
Parameter Stock(f,c,p)
/Active.Home.p12=42
  Reserve.Home.p24=34/;
```

12. *Force Contribution Levels (Effectiveness when deployed):*

```
Parameter Contrib(f,fc,p)
/Active.fc0.p01*p12=1

...

ReserveIneff2.fc5.p03*p24 = 0/;
```

13. *Number of Bases for Each Training Type:*

```
Parameter Bases(t)
/tr1=4

...

tr5=0/;
```

14. *Training Needs When Deployed Indicated Number of Periods:*

```
Parameter Train(f, t, p)
/Active.tr1.p01*p12=0

...

ReserveIneff2.tr5.p01*p24=0/
```

15. *Number of Iterations to Perform:*

```
Scalar numiter /1000/;
```

16. *Contingencies to Include:*

```
Set cont /StochHM11/;
```

17. *Contingency States:*

```
Set state /Peace, War, Sabil/;
```

18. *Initial State:*

```
set curstate(cont,state);
curstate('SotchHM11','Peace') = yes;
```

19. *Transition Probabilities:*

```
Parameter Trans(cont,state,nstate)
/StochHM11.Peace.Peace = 0.975
  StochHM11.War.Peace=0.2

  ...

  StochHM11.Peace.Sabil=0/;
```

20. *Force Requirements by Contingency State and Contribution Type:*

```
Parameter Reqmt(cont,state,fc)
/StochHM11.Sabil.fc0=11

...

StochHM11.Peace.fc5=0/;
```

GAMS Core Program

The GAMS core program takes the information from the GAMS input template and solves the resulting integer or linear program. The program minimizes the objective function subject to the constraints that we formalized in this appendix. The results are read into tables in the back-end database and are immediately ready for analysis using the tools mentioned earlier.

The core program, which solves as an integer program, is reproduced below. The linear program version can be obtained by changing the three lines that are displayed in bold in the program below. Comments are shown in italics.

```
$Title SLAM Version C3
$offlisting
$offsymxref
$offsymlist
* fmax(f,c) is the max number of periods for force type f, current state
* c
* pmax is the max of all these
* lmax is the number of look levels
    Parameter fmax(f,c), pmax, lmax;
    fmax(f,c) = smax(p$Util(f,c,'Deployed',p,'L1'), ord(p));
    pmax = smax((f,c), fmax(f,c));
    lmax = card(l);
    Alias (fc, e);
    Parameter Req(l, e);


* Input parameter Contrib is the contribution defined in terms of period
* deployed, counting the upcoming period. (Home/Deployed is not a
* dimension, since Contrib only applies to forces that will be deployed.)


* Parameter Contr converts this to count in terms of the current
* period, to be compatible with the solution variable X. Current Home/
* Deployed status is therefore a dimension. This parameter only applies
* to forces ordered to be/stay deployed, so ordered Home/Deployed is
* not a dimension.

    Parameter Contr(f,c,e,p);
      Contr(f,'Home',e,p)$(ord(p)<=fmax(f,'Home'))  = Contrib(f,e,'p01');
      Contr(f,'Deployed',e,p)$(ord(p)<fmax(f,'Deployed')) = Contrib(f,e,p+1);
      Contr(f,'Deployed',e,p)$(ord(p)=fmax(f,'Deployed')) = Contrib(f,e,p);


* Parameter Tng converts Train in the same way as Contr does to Contrib.
    Parameter Tng(f,t,c,p);
      Tng(f,t,'Home',p)$(ord(p)<=fmax(f,'Home'))        = Train(f,t,'p01');
      Tng(f,t,'Deployed',p)$(ord(p)<fmax(f,'Deployed')) = Train(f,t,p+1);
      Tng(f,t,'Deployed',p)$(ord(p)=fmax(f,'Deployed')) = Train(f,t,p);


* Other miscellaneous parameter setups.
```

```
    loop(l,
     if (ord(l) > 1,
     Util(f,c,'Deployed',p,l)$(ord(p) <= fmax(f,c)) =
             alpha + (beta * Util(f,c,'Deployed',p,l-1));
      Util(f,c,'Deployed',p,l)$(ord(p) <= fmax(f,c)) =
             Util(f,c,'Deployed',p,l-1);
      Util(f,c,'Home',p,l)$(ord(p) <= fmax(f,c)) =
             Util(f,c,'Home',p,l-1);
      );
     );


    Variables
     X(f, c, n, p, l)   Number of units ordered into each state
     S(f, c, p, l)      Number of units starting in each state
     Z                  Total cost to be minimized            ;


    Integer Variable x;
    * Replace with: Positive Variable X; to solve as linear program
    Positive Variable s;
    Parameter toteff(e);

* For loop 1, S is fixed as the input parameter Stock
        S.fx(f,c,p,'L1')$(ord(p) <= fmax(f,c)) = Stock(f,c,p);
        S.l(f,c,p,l)$(ord(p) <= fmax(f,c))      = Stock(f,c,p);

* Tell the solver not to play with p values beyond each max for f and c
        X.fx(f,c,n,p,l)$(ord(p) > fmax(f,c)) = 0;


     Equations
     COST               define objective function
     CHANGE(f,c,l)      count units changing status
     STAYN(f,c,p,l)     count units remaining in status (non-absorbing case)
     STAYA(f,c,p,l)     count units remaining in status (absorbing case)
     TRANSIT(f,c,p,l)   specify transition from starting to ordered states
     MILREQ(l,e)        satisfy required force level
     BASEREQ(t,l)       satisfy required number of bases of each type ;
     COST   .. Z =e= sum((f, c, n, p, l)$(ord(p) <= fmax(f,c)),
                     Util(f,c,n,p,l) * X(f,c,n,p,l))  ;

* Note, assuming there are two possible values of current status,
* for any current status c, c--1 is the other possible value.
        CHANGE(f,c,l)$(ord(l) < lmax)   ..
        S(f, c, 'p01', l+1) =e= sum(p$(ord(p) <= fmax(f,c)), X(f,c--1,c,p,l));

* This handles forces not changing status and not reaching the
* absorbing state
```

```
      STAYN(f,c,p,l)$(ord(l) < lmax and ord(p) < fmax(f,c)-1) ..
      S(f, c, p+1, l+1) =e= X(f,c,c,p,l);


* This handles forces not changing status and reaching or already in the
* absorbing state
      STAYA(f,c,p,l)$(ord(l) < lmax and ord(p) = fmax(f,c)-1) ..
      S(f, c, p+1, l+1) =e= X(f,c,c,p,l) + X(f,c,c,p+1,l);


* Forces ordered into new states must sum to the forces available
      TRANSIT(f,c,p,l)$(ord(p) <= fmax(f,c)) ..
      sum(n, X(f,c,n,p,l)) =e= S(f,c,p,l);
      MILREQ(l,e)  .. sum((f,c,p)$(ord(p) <= fmax(f,c)),
      Contr(f,c,e,p) * X(f,c,'Deployed',p,l) ) =g= Req(l,e);
      BASEREQ(t,l) .. sum((f,c,p)$(ord(p) <= fmax(f,c)),
      Tng(f,t,c,p) * X(f,c,'Deployed',p,l)) =l= Bases(t) ;

      Option MIP = Cplex;
       * Replace with: Option LP=Cplex; to solve as linear program
      Option Solprint = off;
      Option optcr = 0.0;
      Model slamb /all/ ;


* Set up output files
  file sol /c2sol.csv/, nxt /c2nxt.csv/, sta /c2sta.csv/, mod
  /c3mod.csv/;
      sol.pc = 5;
      nxt.pc = 5;
      sta.pc = 5;
      mod.pc = 5;


* Calculate requirements used in lookaheads just once
* Creqs(l,cont,state,e) is requirement for each look level, contingency,
* initial ("real") state (NOT the lookahead state), and contribution


* For each contingency and look-ahead level (beyond L1)
* Ptrans is the transition matrix, raised to 1 more power on each pass
* find maxp = highest probability of next state (nstate) in the current
* power of the transition matrix
* first = first nstate with prob maxp (to disambiguate ties)
* Creqs = requirement for nstate = first
* requirement for the look-ahead is sum of Creqs over the contingencies
      Alias (state,kstate);
      Parameter Ptrans(cont, state,nstate)  Powers of transition matrix ;
      Parameter Creqs(l,cont,state,e);
      Set newstate (cont,nstate)   Values of curstate for next iteration ;
      Scalar i, st, maxp, first, rand, totsec;
```

```
    Set st1(state), st2(state), st3(state), st4(state), st5(state),
xs1(state), xs2(state), xs3(state), xs4(state), xs5(state), cg1(cont),
cg2(cont), cg3(cont), cg4(cont), cg5(cont);
    Parameter xprob(state, state, state, state, state)
    Treqs(l, state, state, state, state, state, e);
    scalar maxprob, totprob, pb1, pb2, pb3, pb4, pb5, y, yy;

* cg1-cg5 are singleton sets, each containing the single contingency 1-5
* Note: GAMS lacks the concept of a variable that can be set to a set
* element, so it is necessary to sum over singleton sets instead.
    cg1(cont)  = yes$(ord(cont)=1);
    cg2(cont)  = yes$(ord(cont)=2);
    cg3(cont)  = yes$(ord(cont)=3);
    cg4(cont)  = yes$(ord(cont)=4);
    cg5(cont)  = yes$(ord(cont)=5);

* st1-st5 are subsets of states reachable in contingencies 1-5
    st1(nstate) = yes$(sum((cont,state)$(ord(cont)=1),
      Trans(cont,state,nstate)) > 0);
    st2(nstate) = yes$(sum((cont,state)$(ord(cont)=2),
      Trans(cont,state,nstate)) > 0);
    st3(nstate) = yes$(sum((cont,state)$(ord(cont)=3),
      Trans(cont,state,nstate)) > 0);
    st4(nstate) = yes$(sum((cont,state)$(ord(cont)=4),
      Trans(cont,state,nstate)) > 0);
    st5(nstate) = yes$(sum((cont,state)$(ord(cont)=5),
      Trans(cont,state,nstate)) > 0);

* If any of st1-st5 is empty, add the first state (presumably "Peace")
* This is needed because all loops over st1-st5 must have at least one
* pass
    if ((card(st1) < 1), st1(state) = yes$(ord(state) = 1); );
    if ((card(st2) < 1), st2(state) = yes$(ord(state) = 1); );
    if ((card(st3) < 1), st3(state) = yes$(ord(state) = 1); );
    if ((card(st4) < 1), st4(state) = yes$(ord(state) = 1); );
    if ((card(st5) < 1), st5(state) = yes$(ord(state) = 1); );
    Alias(st1, ns1);
    Alias(st2, ns2);
    Alias(st3, ns3);
    Alias(st4, ns4);
    Alias(st5, ns5);

* Initialize Creqs, Treqs, and Ptrans for level 1 ("real world")
    Creqs('L1',cont,state,e)  = Reqmt(cont,state,e);
    Ptrans(cont,state,nstate) = Trans(cont,state,nstate);
```

```
    Treqs('L1',st1,st2,st3,st4,st5,e)   = sum((cg1,cg2,cg3,cg4,cg5),
Reqmt(cg1,st1,e) + Reqmt(cg2,st2,e) + Reqmt(cg3,st3,e) + Reqmt(cg4,st4,e) +
Reqmt(cg5,st5,e) );


* Remaining logic is for lookheads, level >=2
      loop(l$(ord(l) > 1),


* THIS IS THE "MOST-LIKELY" LOOKAHEAD REQUIREMENT METHOD
    if ((tolerance < .01 or tolerance > .99),
        loop(cont,
          loop(state,
             maxp  = smax(nstate, Ptrans(cont,state,nstate) );
             first = smin(nstate$(Ptrans(cont,state,nstate) >= maxp),
                    ord(nstate));
             Creqs(l,cont,state,e) =
                 sum(nstate$(ord(nstate)=first), Reqmt(cont,nstate,e));
    l,cont,state,nstate), ', '; );
              );
          );


* THIS IS THE "TOLERANCE" LOOKAHEAD REQUIREMENT METHOD
    else
* Determine requirements for every combination of initial states st1-st5
    loop(st1, loop(st2, loop(st3, loop(st4, loop(st5,


* Initialize probability xprob of transitioning to states ns1-ns5
* Set maxprob to max value in xprob
* Set xs1-xs5 to singleton sets containing the states with maxprob
    maxprob = 0;
            loop(ns1,
              if ((card(ns1) < 2), pb1 = 1;
              else pb1 = sum(cg1, Ptrans(cg1,st1,ns1));  );


            loop(ns2,
              if ((card(ns2) < 2), pb2 = 1;
              else pb2 = sum(cg2, Ptrans(cg2,st2,ns2));  );


            loop(ns3,
              if ((card(ns3) < 2), pb3 = 1;
              else pb3 = sum(cg3, Ptrans(cg3,st3,ns3));  );


            loop(ns4,
              if ((card(ns4) < 2), pb4 = 1;
              else pb4 = sum(cg4, Ptrans(cg4,st4,ns4));  );
```

```
        loop(ns5,
                            if ((card(ns5) < 2), pb5 = 1;
                            else pb5 = sum(cg5, Ptrans(cg5,st5,ns5));   );


                            y = pb1 * pb2 * pb3 * pb4 * pb5;
                            xprob(ns1, ns2, ns3, ns4, ns5) = y;


                            if ((y > maxprob),
                              maxprob = y;
                              xs1(state) = no; xs2(state) = no; xs3(state) = no;
                              xs4(state) = no; xs5(state) = no;
                              xs1(ns1) = yes; xs2(ns2) = yes; xs3(ns3) = yes;
                              xs4(ns4) = yes; xs5(ns5) = yes;
                              );
                          );
                      );
                  );
              );
```

* *Initially, set Treqs for st1-st5 to the requirement for the most*
* *probable combination of next states (xs1-xs2)*

```
        Treqs(1,st1,st2,st3,st4,st5,e) = sum((cg1,xs1), Reqmt(cg1,xs1,e)) +
        sum((cg2,xs2), Reqmt(cg2,xs2,e)) + sum((cg3,xs3), Reqmt(cg3,xs3,e)) +
sum((cg4,xs4), Reqmt(cg4,xs4,e)) + sum((cg5,xs5), Reqmt(cg5,xs5,e));
        totprob = maxprob;
```

* *Repeat finding next most likely combinations of next states xs1-5*
* *until tolerance reached or exceeded.*

```
        while ((totprob < tolerance),
```

* *If tolerance not yet reached, eliminate xs1-xs5 from consideration*

```
        xprob(xs1,xs2,xs3,xs4,xs5) = 0;
        maxprob = 0;
```

* *Find a new most-likely state combination xs1-xs5*

```
        loop(ns1, loop(ns2, loop(ns3, loop(ns4, loop(ns5,
        y = xprob(ns1,ns2,ns3,ns4,ns5);
                      if ((y > maxprob),
                        maxprob = y;
                        xs1(state) = no; xs2(state) = no; xs3(state) = no;
                        xs4(state) = no; xs5(state) = no;
                        xs1(ns1) = yes; xs2(ns2) = yes; xs3(ns3) = yes;
                        xs4(ns4) = yes; xs5(ns5) = yes;
                        );
                    ); ); ); ); );
```

```
* Treqs is the max requirement for all xs1-xs5 combinations checked
    loop(e,
         y = Treqs(l,st1,st2,st3,st4,st5,e);
    yy = sum((cg1,xs1), Reqmt(cg1,xs1,e)) +  sum((cg2,xs2),
         Reqmt(cg2,xs2,e)) + sum((cg3,xs3), Reqmt(cg3,xs3,e)) +
             sum((cg4,xs4), Reqmt(cg4,xs4,e)) + sum((cg5,xs5),
         Reqmt(cg5,xs5,e));
         Treqs(l,st1,st2,st3,st4,st5,e) = max(y,yy);
         );

    totprob = totprob + maxprob;
               );
          ); ); ); ); );
        );


* For both methods, update Ptrans for next value of l (if any)
        if ((ord(l) < lmax),
           loop(cont,
              Ptrans(cont,state,nstate) = sum(kstate,
                  Ptrans(cont,state,kstate)  *
    Trans(cont,kstate,nstate));
              );
           );
        );
    totsec = 0;


* MONTE CARLO LOOP BEGINS HERE

    for (i = 1 to numiter,
      if ((tolerance < .01 or tolerance > .99),
       Req(l,e)  = sum((cont,state)$curstate(cont,state),
       Creqs(l,cont,state,e));
      else

* Note: Each of cg1-5 is a singleton and has only one st1-5 for which
* curstate is true, so this is really a "sum" of one Treqs value.
    Req(l,e) = sum((cg1,cg2,cg3,cg4,cg5,st1,st2,st3,st4,st5)$(
curstate(cg1,st1) and curstate(cg2,st2) and curstate(cg3,st3)
    and curstate(cg4,st4) and curstate(cg5,st5) ),
    Treqs(l,st1,st2,st3,st4,st5,e)  );
    );
    toteff(e) = sum((f,c,p)$(ord(p)<=fmax(f,c)), S.l(f,c,p,'L1') *
Contr(f,c,e,p) );


* Solve the LP only if sufficient forces exist;
    if (smin(e, Req('L1',e) - toteff(e)) > 0, st = 4;
```

```
     else
         Solve slamb using MIP minimizing Z ;
          * Replace with: Solve slamb using LP minimizing Z; to solve as
            linear program
         st = slamb.modelstat ;
      );


* If solver terminates on a timeout, this is the last iteration
     if (slamb.solvestat = 3, i = numiter + 1);
          put sol;
     if (st < 2,
        loop((f,c,n,p)$(ord(p) <= fmax(f,c)),
        put f.tl c.tl p.tl n.tl X.l(f,c,n,p,'L1') i /);
     else
        loop((f,c,n,p)$(ord(p) <= fmax(f,c)),
        put f.tl c.tl p.tl n.tl "-1" i /);
        );


* Loop 2 and beyond are lookaheads, so force levels at the start
* of loop 2 are the actual levels resulting from the optimization
     put nxt;
     loop((f,c,p)$(ord(p) <= fmax(f,c)), put f.tl c.tl p.tl
S.l(f,c,p,'L2') i /);
     put sta;
     loop((cont,state)$curstate(cont,state), put cont.tl state.tl i /);


* Set newstate for each contingency, given its current state
* rand is a random number between 0 and 1. If rand is <= prob of 1st
* state, newstate will be the 1st state. Otherwise is rand <= prob of
* 1st+2nd state, newstate will be the 2nd state and so on. Note that
* code subtracts probs from rand rather than doing a running sum. This
* is equivalent but simpler.
       if (i < numiter,
          newstate(cont,nstate) = no;
          loop((cont,state)$curstate(cont,state),
             rand = uniform(0, 1);
             loop(nstate$(rand > 0),
                if (rand <= Trans(cont,state,nstate),
                    newstate(cont,nstate) = yes; );
                rand = rand - Trans(cont,state,nstate);
             );
          );


* Now set curstate for the next iteration
     curstate(cont,nstate) = newstate(cont,nstate);
     display newstate;
```

```
* Update the initial disposition of forces for next pass
    S.fx(f,c,p,'L1')$(ord(p) <= fmax(f,c)) = S.l(f,c,p,'L2');
    );
    totsec = totsec + slamb.resusd;

* THIS IS THE END OF THE MONTE CARLO LOOP
    );
    put mod;
    put slamb.numequ totsec slamb.numvar slamb.numnz / ;
```

Creating Assignment Rules

The analyses contained in the body of this document use a variety of different assignment rules. The simplest of them, *ActivesOnly*, is used in Chapters Two and Three of this report. In this assignment rule, only active component forces are utilized. For this and similar assignment rules that include only forces that are fully effective in all periods, determining the assignment rankings is trivial. All an analyst needs to do in these cases is to rank all of the force-status-periods (Act-Home-1 through Act-Home-Max, and Act-Dep-1 through Act-Dep-Max) in the order in which he or she would like forces to be deployed.

The process of choosing an assignment rule is more complicated for forces that are not fully effective in all periods. This appendix describes how to create these types of assignment rules. Using the assignment rules created in Chapters Four and Five of this report as examples, this appendix describes how different types of assignment rules can be created for different needs.

D.1. Base Active/Reserve Assignment Rule

We begin by describing the methodology used to create *BaseActiveReserve*, the assignment rule discussed in Section 4.4 and used in the majority of the analyses in this report. Each assignment rule is created with a goal in mind. This assignment rule was created with the goals of (1) limiting active deployments to 1 year, (2) limiting reserve mobilizations to 1 year (0.5 year deployed overseas, 0.5 year of training), and (3) limiting reserve utilization to 1 year in 6. To meet any larger requirements, the factor that we wish to vary is active component time at home. This means that once actives are deployed 1 year in every 3 and reserves are deployed 1 year in every 6, the program should begin deploying actives with shorter dwell times.

To determine the appropriate coefficients for the objective function (the "disutilities"), we need to use a spreadsheet that can calculate average disutility per effective period (described in Section D.1.1). The SLAM distribution package comes with a Microsoft Excel spreadsheet (CreateAssignmentRule.xls) that does exactly this. This spreadsheet can be opened by clicking on the *Create Rule in Excel* button on the Advanced Assignment Rule subform, as shown in Figure D.1. We suggest that the user not make any direct changes to the CreateAssignmentRule.xls file, since this file serves as the template for Access to create the correct spreadsheet and changing it could corrupt this process.

Figure D.1
Create [Assignment] Rule in Excel Button

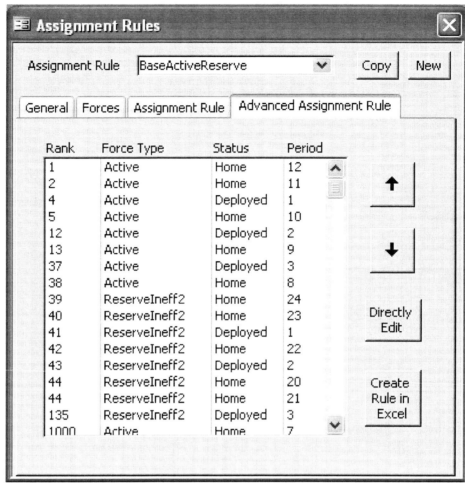

RAND *TR433-D.1*

Clicking on the *Create Rule in Excel* button will open an Excel spreadsheet that is automatically saved in the working directory and named by concatenating the name of the backend database and the name of the assignment rule. The spreadsheet will look similar to that shown in Figure D.2.

Figure D.2
Spreadsheet for Defining Assignment Rules

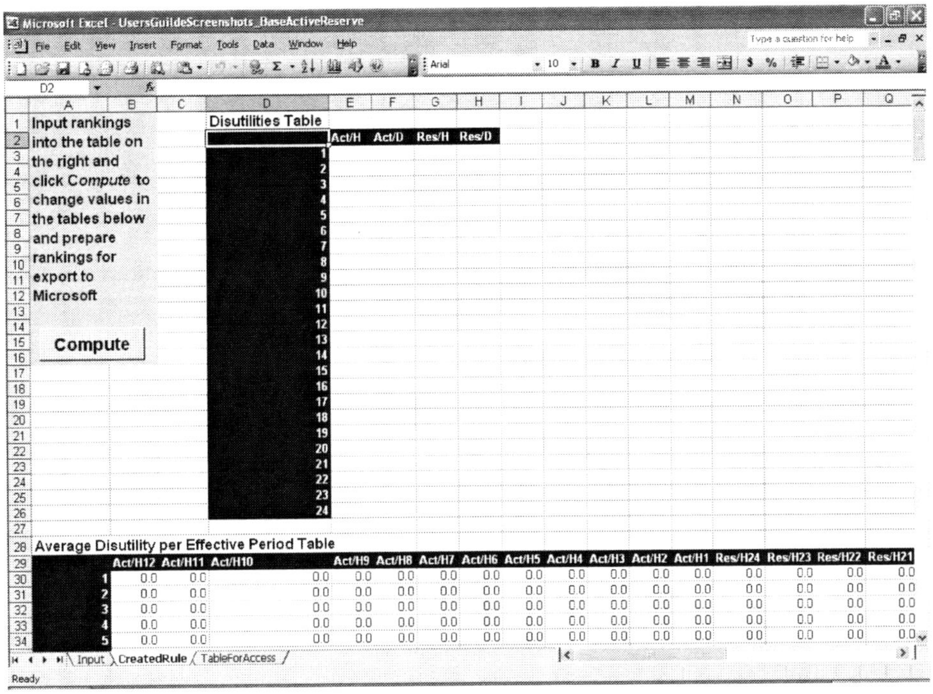

RAND *TR433-D.2*

The spreadsheet that the SLAM program creates is based on the number of forces included in the assignment rule, the maximum number of periods home and deployed for each force, and the number of periods of deployment in which each force is effective. This is all the information needed to create the spreadsheet.

The important calculation in this spreadsheet is the average disutility per effective period. The average disutility per effective period is defined as the sum of the disutilities of deploying a force for each period up to the current deployed period, divided by the number of those periods in which the force is effective. This calculation is important because of the structure of the linear program. Since simulations with ineffective forces use look-ahead periods, in any given period the LP tries to minimize the total disutility across all look-ahead periods. This means that, when considering whether to deploy a unit today for 4 periods, the LP will determine which unit to send based on the average disutility of deploying a unit for the next 4 periods, and not on the direct disutility of deploying a unit for just the next period. For this reason, we must focus our attention on average disutility per effective period rather than just the disutilities.

There are three separate tables needed to calculate average disutility per effective period: the Average Disutility per Effective Period table, the Disutility of Deployment table, and the Disutilities table. Each of these tables is included in the spreadsheet created by clicking on the *Create Rule in Excel* button, as shown in Figure D.3. The Disutilities table is at the top of the spreadsheet, with the Average Disutility per Effective Period table below it and the Disutility of Deployment at the bottom.

Figure D.3
Full Spreadsheet for Defining Assignment Rules

Input rankings into the table on the right and click C *ompute* to change values in the tables below and prepare rankings for export to Microsoft Access.

Compute

Disutilities Table

	Act/H	Act/D	Res/H	Res/D
1	24576000	4	999999999	41
2	6144000	12	999999999	43
3	1536000	37	999999999	135
4	384000	999999999	999999999	405
5	96000	999999999	999999999	1305
6	24000	999999999	999999999	999999999
7	6000	999999999	999999999	999999999
8	38	999999999	999999999	999999999
9	13	999999999	999999999	999999999
10	5	999999999	999999999	999999999
11	2	999999999	999999999	999999999
12	1	999999999	999999999	999999999
13			999999999	999999999
14			999999999	999999999
15			999999999	999999999
16			999999999	999999999
17			999999999	999999999
18			1306	999999999
19			406	999999999
20			136	999999999
21			44	999999999
22			42	999999999
23			40	999999999
24			39	999999999

Average Disutility per Effective Period Table

	Act/H12	Act/H11	Act/H10	Act/H9	Act/H8	Act/H7	Act/H6	Act/H5	Act/H4
1	1.0	2.0	5.0	13.0	38.0	6000.0	24000.0	96000.0	384000.0
2	2.5	3.0	4.5	8.5	21.0	3002.0	12002.0	48002.0	192002.0
3	5.7	6.0	7.0	9.7	18.0	2005.3	8005.3	32005.3	128005.3
4	13.5	13.8	14.5	16.5	22.8	1513.3	6013.3	24013.3	36013.3
5	200000010.6	200000010.8	200000011.4	200000013.0	200000018.0	200001210.4	200004810.4	200019210.4	200076810.4
6	333333342.0	333333342.2	########	333333344.0	333333348.2	333334341.8	333337341.8	333349341.8	333397341.8
7	428571435.9	428571436.0	428571436.4	428571437.6	428571441.1	428572232.9	428574864.3	428585150.0	428626292.9
8	500000006.3	500000006.4	########	500000007.8	500000010.9	500000756.1	500003006.1	500012006.1	500048006.1
9	555555561.0	555555561.1	555555561.4	555555562.3	555555565.1	555556227.6	555558227.6	555566227.6	555538227.6
10	600000004.8	600000004.9	########	600000006.0	600000008.5	600000604.7	600002404.7	600003604.7	600038404.7
11	636363640.6	636363640.7	########	636363641.7	636363644.0	636364186.0	636365822.4	636372367.8	636398543.6
12	666666670.5	666666670.6	########	666666671.5	666666673.6	666667170.4	666668670.4	666674670.4	666638670.4
13									
14									
15									
16									
17									
18									
19									
20									
21									
22									
23									
24									

Disutility of Deployment Table

	Act/H12	Act/H11	Act/H10	Act/H9	Act/H8	Act/H7	Act/H6	Act/H5	Act/H4
1	1	2	5	13	38	6000	24000	96000	384000
2	4	4	4	4	4	4	4	4	4

RAND *TR433-D.3*

D.1.1. Calculating Average Disutility per Effective Period

The most important calculation for defining an assignment rule is the average disutility per effective period. The calculation of the average disutility per effective period depends on the number of periods for which a unit is deployed, the disutilities associated with deploying that unit, and the number of periods of deployment in which a force is ineffective. We describe this calculation for the active and reserve forces used in the analyses in this report.

Active Component Average Disutility per Effective Period. For the active component, this calculation is simply the sum of the disutilities of being deployed up to a given number of periods, divided by the number of periods deployed (since actives are fully effective in all periods they are deployed).

If we define d as an index of the number of periods deployed and h as an index of the number of periods at home before a unit was deployed, we can represent the disutility of any deployment as U_{hd}. The equation below shows how to calculate the average disutility of deployment, which we define as AU_{HD} for an active unit deployed for D periods after being at home for H periods:

$$AU_{HD}^{\text{Active}} = \frac{\sum\limits_{i=1}^{D} U_{Hi}}{D}$$

Reserve Component Average Disutility per Effective Period. For the reserve component (and any other force that is not effective in all periods), the calculation of average disutility per effective period is different from the active component. For the reserves, we still sum the disutilities of being deployed up to a given number of periods, but we divide by the total number of deployed periods minus the number of periods in which the force is ineffective. The reserves used in the analyses in this report are assumed to be ineffective for the first two periods of deployment. Therefore, we will divide by $D - 2$, as shown in the equation below. We set the average disutility per effective period for ineffective periods of deployment (periods 1 and 2 in this case) equal to 0, as these values are ignored in later calculations.

$$\forall D > 2 : AU_{HD}^{\text{Reserve}} = \frac{\sum\limits_{i=1}^{D} U_{Hi}}{D - 2}$$

$$\forall D \leq 2 : AU_{HD}^{\text{Reserve}} = 0$$

In other analyses, other assumptions may be made about force train-up time. A more generic version of the equation above will depend on the force effectiveness parameter. If we define IE as the number of periods that a force is ineffective when initially deployed, then a more general version of the average disutility per effective period is

$$\forall D > IE : AU_{HD} = \frac{\sum\limits_{i=1}^{D} U_{Hi}}{D - IE}$$

$$\forall D \leq IE : AU_{HD} = 0$$

The three tables described in the next three sections help to calculate average disutility per effective period by capturing the disutilities entered by the user (Disutilities table), using these disutilities to calculate the disutility of deploying a unit from any period for any number of periods (Disutility of Deployment table), and using the Disutility of Deployment table to calculate average disutility per effective period.

D.1.2. Disutilities Table

The Disutilities table contains the objective function coefficients that will be assigned as "disutilities" in the SLAM program. This table includes every possible state for every force (home or deployed X number of periods). For instance, in the analyses performed in Chapter Four, there are 24 possible states for the actives (12 home and 12 deployed) and 48 possible states for the reserves (24 home and 24 deployed). Figure D.4 shows the Disutilities table that defines the *BaseActiveReserve* assignment rule used in many of the analyses in this document. The columns are defined by force-state (active-home, active-deployed, reserve-home, reserve-deployed). The rows are defined by the number of periods *currently* in the corresponding state. For instance, *Act/H-1* is an active unit currently at home for 1 period and *Res/D-6* is a reserve unit currently deployed for 6 periods. The disutilities in this table are the objective function coefficients assigned to each decision variable in the linear program in which deployed is the state in the *next* period (all decision variables in which home is the state in the next period have disutilities hard-coded as 0).

Figure D.4
Disutilities Table

Disutilities Table

	Act/H	Act/D	Res/H	Res/D
1	4E+06	4	1E+09	41
2	1E+06	12	1E+09	43
3	256000	37	1E+09	135
4	64000	1E+09	1E+09	1E+09
5	16000	1E+09	1E+09	1E+09
6	4000	1E+09	1E+09	1E+09
7	1000	1E+09	1E+09	1E+09
8	38	1E+09	1E+09	1E+09
9	13	1E+09	1E+09	1E+09
10	5	1E+09	1E+09	1E+09
11	2	1E+09	1E+09	1E+09
12	1	1E+09	1E+09	1E+09
13			1E+09	1E+09
14			1E+09	1E+09
15			1E+09	1E+09
16			1E+09	1E+09
17			1E+09	1E+09
18			1E+09	1E+09
19			1E+09	1E+09
20			136	1E+09
21			44	1E+09
22			42	1E+09
23			40	1E+09
24			39	1E+09

RAND *TR433-D.4*

The user will modify the disutilities in this table to obtain a desired deployment ordering. The disutilities entered by the user are used to fill the Disutility of Deployment table, which is then used to calculate the average disutility per effective period. It is important to note that neither the Disutility of Deployment table nor the Average Disutility per Effective Period table updates values until the *Compute* button (located in the top left corner of the Excel spreadsheet) is selected by the user. Therefore, after making changes to the Disutilities table, the user should always click the *Compute* button to update all of the tables in the spreadsheet. Selecting *Compute* also ensures that the disutilities are prepared for export back into the Access database.

If we were using only active-duty forces, this whole process would be unnecessary because we could just assign the lowest disutility to the force-status-period we would like to deploy first and increment the disutilities until each force-status-period has a ranking. However, when there are forces that are not fully effective in all periods, it is not the direct disutility we need to be concerned with, but instead the average disutility per effective period. For this reason, we must use the Disutility of Deployment table and the Average Disutility per Effective Period table, as described below.

D.1.3. Average Disutility per Effective Period Table

This table is the key component to designing an assignment rule with forces that are not fully effective in all periods. This table has a column for each period that each force could be at home and a row for each period of deployment. Each cell in this table uses the disutilities from the Disutilities table to calculate the *average disutility* per *effective* period, as described earlier.

This table includes a column for each force-period at home and a row for each possible period of deployment. In this example, there are 12 columns for the actives and 24 columns for the reserves, for a total of 36 columns. Since the maximum number of periods home and deployed is the same for each force, there are also 12 rows for the actives and 24 rows for the reserves. Figure D.5 shows the first six columns of the Average Disutility per Effective Period table for the *BaseActiveReserve* assignment rule. Other portions of this table will be reproduced in later sections.

Figure D.5
Average Disutility per Effective Period Table

Average Disutility per Effective Period Table						
	Act/H12	Act/H11	Act/H10	Act/H9	Act/H8	Act/H7
1	1.0	2.0	5.0	13.0	38.0	6000.0
2	2.5	3.0	4.5	8.5	21.0	3002.0
3	5.7	6.0	7.0	9.7	18.0	2005.3
4	13.5	13.8	14.5	16.5	22.8	1513.3
5	200000010.6	200000010.8	200000011.4	200000013.0	200000018.0	200001210.4
6	333333342.0	333333342.2	333333342.7	333333344.0	333333348.2	333334341.8
7	428571435.9	428571436.0	428571436.4	428571437.6	428571441.1	428572292.9
8	500000006.3	500000006.4	500000006.8	500000007.8	500000010.9	500000756.1
9	555555561.0	555555561.1	555555561.4	555555562.3	555555565.1	555556227.6
10	600000004.8	600000004.9	600000005.2	600000006.0	600000008.5	600000604.7
11	636363640.6	636363640.7	636363641.0	636363641.7	636363644.0	636364186.0
12	666666670.5	666666670.6	666666670.8	666666671.5	666666673.6	666667170.4

RAND *TR433-D.5*

The cell values in this table are calculated using the equations described in Section D.1.1. These calculations are simplified by the organization of the Disutility of Deployment table discussed in the next section.

D.1.4. Disutility of Deployment Table

This table also includes a column for each force-period at home and a row for each possible period of deployment. Figure D.6 displays part of this table, showing only the first six of the 36 total columns.

Figure D.6
Disutility of Deployment Table

Disutility of Deployment Table						
	Act/H12	Act/H11	Act/H10	Act/H9	Act/H8	Act/H7
1	1	2	5	13	38	1000
2	4	4	4	4	4	4
3	12	12	12	12	12	12
4	37	37	37	37	37	37
5	999999999	999999999	999999999	999999999	999999999	999999999
6	999999999	999999999	999999999	999999999	999999999	999999999
7	999999999	999999999	999999999	999999999	999999999	999999999
8	999999999	999999999	999999999	999999999	999999999	999999999
9	999999999	999999999	999999999	999999999	999999999	999999999
10	999999999	999999999	999999999	999999999	999999999	999999999
11	999999999	999999999	999999999	999999999	999999999	999999999
12	999999999	999999999	999999999	999999999	999999999	999999999

RAND *TR433-D.6*

The cell values in this table are calculated using the values in the Disutilities table. In the Disutility of Deployment table, the entries in the first row for every column are equal to the disutility of deploying a unit at home from the number of periods corresponding to the column for one period. This can be verified by comparing the values in the first column of Figure D.4 with the values in the first row of Figure D.6. If one starts at the bottom of the first column of disutilities in Figure D.4 and moves up, these values are exactly the same as the values moving from left to right in the first row in Figure D.6. The values in each cell of row 2 are identical to each other because this is the disutility of deploying a unit that has already been deployed for one period (the same for all units no matter where they start from). This is true for all rows other than the first (2–12 for the actives, 2–24 for the reserves). To summarize, this table provides the direct disutility of deploying a unit for any number of periods from every possible previous state. The user should not need to modify this table. This table is required because it allows the spreadsheet to easily calculate the average disutility per effective period as discussed in the previous sections.

D.1.5. Determining Disutilities

The disutilities are determined by the user based on the desired values in the Average Disutility per Effective Period table. The base assignment rule that we would like to specify in this section limits active deployments to 4 periods and reserve deployments to 4 periods (2 periods training and 2 periods deployed). If more units are needed, the assignment rule should reduce active time at home. To specify this rule, we start by answering a simple question: Which units should be deployed first? For this assignment rule, we specify that active units at home the longest should be deployed first. Therefore we start by assigning the lowest ranking to actives home 12 periods (*Act/H-12*). In this assignment rule, we choose the lowest ranking to be 1, but this choice is arbitrary. Only the relative difference between disutilities matters, not the levels. After we enter a 1 in the cell corresponding to *Act/H-12* in the Disutilities table and clicking the *Compute* button, the disutility of deployment and average disutility per effective period appear, as in Figure D.7.

Figure D.7
Effect of Setting *Act/H-12* Equal to 1

Disutility of Deployment Table	Act/H12	Act/H11	Act/H10	Act/H9	Act/H8		Average Disutility per Effective Period Table	Act/H12	Act/H11	Act/H10	Act/H9	Act/H8
1	1						1	1.0	0.0	0.0	0.0	0.0
2							2	0.5	0.0	0.0	0.0	0.0
3							3	0.3	0.0	0.0	0.0	0.0
4							4	0.3	0.0	0.0	0.0	0.0
5							5	0.2	0.0	0.0	0.0	0.0
6							6	0.2	0.0	0.0	0.0	0.0
7							7	0.1	0.0	0.0	0.0	0.0
8							8	0.1	0.0	0.0	0.0	0.0
9							9	0.1	0.0	0.0	0.0	0.0
10							10	0.1	0.0	0.0	0.0	0.0
11							11	0.1	0.0	0.0	0.0	0.0
12							12	0.1	0.0	0.0	0.0	0.0

RAND *TR433-D.7*

After assigning an initial ranking, we next determine what units should go next. In the example here, we wish to limit active deployments to 4 periods. From a steady-state perspective, this means that, on average, an active unit would be deployed for 4 periods and home for 8 periods (for a 12-period cycle). Therefore, we concentrate on the part of the Average Disutility per Effective Period table shown in Figure D.8.

Figure D.8
Active Average Disutility per Effective Period

Average Disutility per Effective Period Table

	Act/H12	Act/H11	Act/H10	Act/H9	Act/H8
1	1.0	0.0	0.0	0.0	0.0
2	0.5	0.0	0.0	0.0	0.0
3	0.3	0.0	0.0	0.0	0.0
4	0.3	0.0	0.0	0.0	0.0

RAND *TR433-D.8*

When we assigned forces that were effective in all periods, we simply gave the lowest disutilities to those force-status-periods that we wanted to deploy first. The methodology is similar here, except that, instead of giving the lowest disutility to those force-status-periods we wish to deploy first, we manipulate the disutilities to give the lowest average disutility per effective period to those force-status-periods we wish to deploy first. The important numbers to look at in Figure D.8 and in all other average disutility tables are in the left-most column and the top row (shown in bold above). The analyst will target changes in these cells when entering disutilities into the Disutilities table.

Earlier, we decided that, if there are active forces at home for 12 periods, we want to deploy them first. We therefore assigned *Act/H-12* a value of 1. Next, we would like to ensure that an active unit home 11 periods (deployed 1 period in the steady state) will be deployed. We therefore set *Act/H-11* equal to 2. After we enter a 2 in the cell corresponding to *Act/H-11* in the Disutilities table and click on the *Compute* button, the disutility of deployment and average disutility per effective period appear, as in Figure D.9.

Figure D.9
Effect of Setting *Act/H-11* Equal to 2

Disutility of Deployment Table	Act/H12	Act/H11	Act/H10	Act/H9	Act/H8		Average Disutility per Effective Period Table	Act/H12	Act/H11	Act/H10	Act/H9	Act/H8
1	1	2					1	1.0	2.0	0.0	0.0	0.0
2							2	0.5	1.0	0.0	0.0	0.0
3							3	0.3	0.7	0.0	0.0	0.0
4							4	0.3	0.5	0.0	0.0	0.0

RAND *TR433-D.9*

Next, we want to deploy active units 2 periods, starting with those at home 12 periods. To do this, we must set the disutility of being deployed from *Act/D-1* so that the average disutility of being deployed 2 periods from *Act/H-12 (Act/H-12:D2)* is larger than that of being deployed 1 period from *Act/H-11 (Act/H-11:D1)*. If we were to follow the methodology used for forces effective in all periods of deployment, we would simply set *Act/D-1* equal to 3. However, if we do this, we find that the average disutility of being deployed two periods from *Act/H-12* is equal to that of being deployed one period from *Act/H-11* (2), as shown in Figure D.10.

Figure D.10
Mistakenly Setting *Act/D-1* Equal to 3

Average Disutility per Effective Period Table	Act/H12	Act/H11	Act/H10	Act/H9	Act/H8
1	1.0	2.0	0.0	0.0	0.0
2	2.0	2.5	1.5	1.5	1.5
3	1.3	1.7	1.0	1.0	1.0
4	1.0	1.3	0.8	0.8	0.8

RAND *TR433-D.10*

Because we want to deploy a unit from at home 12 periods for 2 periods only *after* we deploy a unit from at home 11 periods for 1 period, we must set *Act/D-1* higher than 3, so that *Act/H-12:D2* is greater than *Act/H-11:D1*. To make this the case, we found that setting *Act/D-1* equal to 4 led *Act/H-12:D2* to equal 2.5. We found this disutility relatively quickly due to the fact that simply incrementing the disutility produced an average disutility equal to the value that we desired it to be greater than. Therefore, we could simply increment the disutility by one to obtain the desired result. Other disutilities may take more experimenting to find the correct value. After we set *Act/D-1* equal to 4 and clicking the *Compute* button, the disutility of deployment and average disutility per effective period appear, as in Figure D.11.

Figure D.11
Correctly Setting *Act/D-1* Equal to 4

Disutility of Deployment Table	Act/H12	Act/H11	Act/H10	Act/H9	Act/H8		Average Disutility per Effective Period Table	Act/H12	Act/H11	Act/H10	Act/H9	Act/H8
1	1	2					1	1.0	2.0	0.0	0.0	0.0
2	4	4	4	4	4		2	2.5	3.0	2.0	2.0	2.0
3							3	1.7	2.0	1.3	1.3	1.3
4							4	1.3	1.5	1.0	1.0	1.0

RAND *TR433-D.11*

In the steady state, units deployed for 2 periods are home for 10 periods. Therefore, we next wish to deploy units home for 10 periods. To do this, we increment the disutilities by setting *Act/H-5* equal to 5 (1 greater than *Act/D-1*). It is important that the disutilities be incremented, even if one is not utilizing the average disutility table to determine disutilities. We have found that if this convention is not followed, erroneous results can occur. Setting *Act/H-12* equal to 5 and clicking the *Compute* button results in the Disutility of Deployment and Average Disutility per Effective Period tables shown in Figure D.12.

Figure D.12
Setting *Act/H-10* Equal to 5

Disutility of Deployment Table						Average Disutility per Effective Period Table					
	Act/H12	Act/H11	Act/H10	Act/H9	Act/H8		Act/H12	Act/H11	Act/H10	Act/H9	Act/H8
1	1	2	5			1	1.0	2.0	5.0	0.0	0.0
2	4	4	4	4	4	2	2.5	3.0	4.5	2.0	2.0
3						3	1.7	2.0	3.0	1.3	1.3
4						4	1.3	1.5	2.3	1.0	1.0

RAND *TR433-D.12*

The next step is to deploy an active unit for 3 periods (9 periods at home in the steady state). To do this, we need to set *Act/D-2* in exactly the same fashion as we set *Act/D-1*. We would like to set *Act/D-2* so that the average disutility of deploying an active unit for 3 periods from at home 12 periods is larger than the average disutility of deploying an active unit 1 period from at home 10 periods (which we already set to 5 by defining *Act/H-10* to be 5). After some experimentation, we found that setting *Act/D-2* equal to 12 causes the average disutility of being deployed 3 periods from home 12 periods to be 5.7, larger than that of deploying an active unit for 1 period from home 10 periods. We can see this result in the highlighted cells shown in Figure D.13.

Figure D.13
Setting *Act/D-2* Equal to 12

Disutility of Deployment Table						Average Disutility per Effective Period Table					
	Act/H12	Act/H11	Act/H10	Act/H9	Act/H8		Act/H12	Act/H11	Act/H10	Act/H9	Act/H8
1	1	2	5			1	1.0	2.0	5.0	0.0	0.0
2	4	4	4	4	4	2	2.5	3.0	4.5	2.0	2.0
3	12	12	12	12	12	3	5.7	6.0	7.0	5.3	5.3
4						4	4.3	4.5	5.3	4.0	4.0

RAND *TR433-D.13*

Next, we again consider the steady state in which a unit deployed 3 periods will be at home for 9 periods. Therefore, we increment the disutilities so that *Act/H-9* equals 13. This results in the disutilities shown in Figure D.14.

Figure D.14
Setting *Act/H-9* Equal to 13

Disutility of Deployment Table						Average Disutility per Effective Period Table					
	Act/H12	Act/H11	Act/H10	Act/H9	Act/H8		Act/H12	Act/H11	Act/H10	Act/H9	Act/H8
1	1	2	5	13		1	1.0	2.0	5.0	13.0	0.0
2	4	4	4	4	4	2	2.5	3.0	4.5	8.5	2.0
3	12	12	12	12	12	3	5.7	6.0	7.0	9.7	5.3
4						4	4.3	4.5	5.3	7.3	4.0

RAND *TR433-D.14*

The last disutility we set is *Act/D-3*, such that deploying an active unit 4 periods from home 12 periods has a larger average disutility than deploying a unit home 9 periods for 1 period (which we just set equal to 13 by setting *Act/H-9* to 13). After some experimentation, we found that setting *Act/D-3* equal to 37 leads the average disutility of deploying a unit 4 periods from at home 12 periods to equal 13.5, just larger than 13, as shown in Figure D.15.

Figure D.15
Setting *Act/D-3* Equal to 37

Disutility of Deployment Table

	Act/H12	Act/H11	Act/H10	Act/H9	Act/H8
1	1	2	5	13	
2	4	4	4	4	4
3	12	12	12	12	12
4	37	37	37	37	37

Average Disutility per Effective Period Table

	Act/H12	Act/H11	Act/H10	Act/H9	Act/H8
1	1.0	2.0	5.0	13.0	0.0
2	2.5	3.0	4.5	8.5	2.0
3	5.7	6.0	7.0	9.7	5.3
4	13.5	13.8	14.5	16.5	13.3

RAND *TR433-D.15*

Lastly, we set *Act/H-8* equal to 38 by following the incrementing rule. After clicking the *Compute* button, we see the Disutility of Deployment and Average Disutility per Effective Period tables shown in Figure D.16.

Figure D.16
Setting *Act/H-8* Equal to 38

Disutility of Deployment Table

	Act/H12	Act/H11	Act/H10	Act/H9	Act/H8
1	1	2	5	13	38
2	4	4	4	4	4
3	12	12	12	12	12
4	37	37	37	37	37

Average Disutility per Effective Period Table

	Act/H12	Act/H11	Act/H10	Act/H9	Act/H8
1	1.0	2.0	5.0	13.0	38.0
2	2.5	3.0	4.5	8.5	21.0
3	5.7	6.0	7.0	9.7	18.0
4	13.5	13.8	14.5	16.5	22.8

RAND *TR433-D.16*

We have now completed the assignment of active forces home for 8 periods or longer for 4-period deployments. The next step in creating this assignment rule is to deploy the reserves. We therefore move to the part of the average disutility per effective period for the reserves, shown in Figure D.17.

Figure D.17
Reserve Average Disutility per Effective Period

Res/H24	Res/H23	Res/H22	Res/H21	Res/H20
0.0	0.0	0.0	0.0	0.0
0.0	0.0	0.0	0.0	0.0
0.0	0.0	0.0	0.0	0.0
0.0	0.0	0.0	0.0	0.0

RAND *TR433-D.17*

The first thing to note about Figure D.17 is that the first two rows will always have zeroes. We ignore these two rows because these are periods of deployment in which the reserves provide no effective forces. After deploying active units for 4 quarters, we want to deploy reserve units for 4 quarters (2 quarters effective). We begin by incrementing the disutilities for the cells that do not provide effective forces, so that we can focus on those periods in which effective forces are provided. We first set *Res/H-24* equal to 39, then *Res/H-23* equal to 40. Following the same steady-state ordering used for the actives, we next deploy a reserve unit for 2 periods by setting *Res/D-1* equal to 41 (again simply incrementing the disutilities). We then set *Res/H-22* equal to 42, *Res/D-2* equal to 43, and *Res/H-21* equal to 44. After clicking on the *Compute* button, we see the tables shown in Figure D.18.

Figure D.18
Reserve Disutilities After Initial Reserve Increment

Disutility of Deployment Table					Average Disutility per Effective Period Table				
Res/H24	Res/H23	Res/H22	Res/H21	Res/H20	Res/H24	Res/H23	Res/H22	Res/H21	Res/H20
39	40	42	44		0.0	0.0	0.0	0.0	0.0
41	41	41	41	41	0.0	0.0	0.0	0.0	0.0
43	43	43	43	43	123.0	124.0	126.0	128.0	84.0
					61.5	62.0	63.0	64.0	42.0

RAND TR433-D.18

After these initial incremented disutilities, we focus on deploying a reserve unit for 4 periods (2 effective periods). In the steady state, a reserve unit deployed for 4 periods will be at home for 20 periods (given the maximum of 24 periods). Therefore, we want to set *Res/D-3* so that the average disutility of deploying a reserve unit for 4 periods from home 24 periods is larger than that of deploying a reserve unit for 3 periods from at home 21 periods (128). After experimenting with values for *Res/D-3*, we found that setting *Res/D-3* equal to 135 results in the average disutility of being deployed 4 periods from home 24 periods to be 129, as shown in Figure D.19.

Figure D.19
Setting *Res/D-3* Equal to 135

Disutility of Deployment Table					Average Disutility per Effective Period Table				
Res/H24	Res/H23	Res/H22	Res/H21	Res/H20	Res/H24	Res/H23	Res/H22	Res/H21	Res/H20
39	40	42	44		0.0	0.0	0.0	0.0	0.0
41	41	41	41	41	0.0	0.0	0.0	0.0	0.0
43	43	43	43	43	123.0	124.0	126.0	**128.0**	84.0
135	135	135	135	135	**129.0**	129.5	130.5	131.5	109.5

RAND TR433-D.19

The last step in setting rankings for the reserves is to increment the disutility by setting *Res/H-20* equal to 136. This results in the tables shown in Figure D.20.

Figure D.20
Setting *Res/H-20* Equal to 136

Disutility of Deployment Table					Average Disutility per Effective Period Table				
Res/H24	Res/H23	Res/H22	Res/H21	Res/H20	Res/H24	Res/H23	Res/H22	Res/H21	Res/H20
39	40	42	44	136	0.0	0.0	0.0	0.0	0.0
41	41	41	41	41	0.0	0.0	0.0	0.0	0.0
43	43	43	43	43	123.0	124.0	126.0	128.0	220.0
135	135	135	135	135	129.0	129.5	130.5	131.5	177.5

RAND TR433-D.20

The final step in specifying this assignment rule is to deploy active units at home for less than 8 periods on 4-period deployments. Again, this is done using the Average Disutility per Effective Period table. Figure D.21 shows the average disutility table for active units home less than 8 periods and deployed for 4 periods.

Figure D.21
Average Disutility per Effective Period for Reducing Active Time at Home

Act/H7	Act/H6	Act/H5	Act/H4	Act/H3	Act/H2	Act/H1
1000.0	4000.0	16000.0	64000.0	256000.0	1024000.0	4096000.0
502.0	2002.0	8002.0	32002.0	128002.0	512002.0	2048002.0
338.7	1338.7	5338.7	21338.7	85338.7	341338.7	1365338.7
263.3	1013.3	4013.3	16013.3	64013.3	256013.3	1024013.3

RAND *TR433-D.21*

The disutilities the user needs to adjust to set these deployments are *Act/H-7* through *Act/H-1*. We would like to set these so that, once the active and reserves have been sent out on 4-period deployments (on a deployment schedule of 1 year in 3 for actives and 1 year in 6 for reserves), the assignment rule will pull active units at home for shorter periods of time to meet any further requirements. We begin by setting *Act/H-7* so that the average disutility of deploying an active unit 4 periods from at home 7 periods is greater than the largest average disutility from the reserve deployment box (in this case, that is *Res/H-20:D3*, with an average disutility of 220). We set *Act/H-7* equal to 1,000 so that the average disutility of *Act/H-7:D4* equals 263, larger than 220.

Next, we want to ensure that the average disutility of deploying an active unit for 4 periods from at home 6 periods is larger than that of deploying an active unit at home 7 periods for 1 period. After experimenting, we found that setting *Act/H-6* equal to 4,000 caused *Act/H-6:D4* to equal 1,013, slightly greater than *Act/H-7:D-1* (1,000). We repeated the same process to set *Act/H-5* through *Act/H-1*, the results of which are shown in the rest of Figure D.21.

At this point, we have fully specified the deployment ordering. However, there are still unfilled entries in the Disutilities table. When we decided on the assignment rule specified here, we already knew that active/reserve deployments of 4 quarters and bringing actives from home with less than 2 years at home would be able to meet the military requirements considered in this report. Therefore, we filled all of the unfilled cells in the ranking table with a very large number (in this case, 999,999,999) so that forces in those conditions will never be used. In general, an analyst may wish to go further in specifying an assignment rule so that all possible levels of force requirements can be met. For instance, if active time at home was reduced to the extreme, we are left with the options of either increasing deployment lengths (for actives and/or reserves) or decreasing reserve time at home (in similar fashion to what we did for the actives). However, to save time, it is often sufficient to specify rules up to the point at which the analyst knows there are enough rotational forces to meet all possible force requirements. The user might also wish to utilize "phantom" forces to fill larger requirements, as described in Section 6.2.1. The final disutilities for this assignment rule are shown in Figure D.22.

Figure D.22
Disutilities Table for the *BaseActiveReserve* Assignment Rule

Disutilities Table

	Act/H	Act/D	Res/H	Res/D
1	4E+06	4	1E+09	41
2	1E+06	12	1E+09	43
3	256000	37	1E+09	135
4	64000	1E+09	1E+09	1E+09
5	16000	1E+09	1E+09	1E+09
6	4000	1E+09	1E+09	1E+09
7	1000	1E+09	1E+09	1E+09
8	38	1E+09	1E+09	1E+09
9	13	1E+09	1E+09	1E+09
10	5	1E+09	1E+09	1E+09
11	2	1E+09	1E+09	1E+09
12	1	1E+09	1E+09	1E+09
13			1E+09	1E+09
14			1E+09	1E+09
15			1E+09	1E+09
16			1E+09	1E+09
17			1E+09	1E+09
18			1E+09	1E+09
19			1E+09	1E+09
20			136	1E+09
21			44	1E+09
22			42	1E+09
23			40	1E+09
24			39	1E+09

RAND *TR433-D.22*

Once the analyst is satisfied with the assignment rule, the next step is to import the disutilities from the spreadsheet into the SLAM program. This is automatically done by the SLAM program when the user returns to Microsoft Access, with or without closing the spreadsheet. In either case, the SLAM program will present the message box shown in Figure D.23 to make sure that the user wishes to import the disutilities specified on the Excel spreadsheet. We recommend that the user use this method to transfer the disutilities.

Figure D.23
Transfer Assignment Rule Message Box

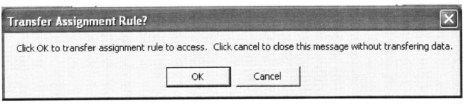

Transfer Assignment Rule?

Click OK to transfer assignment rule to access. Click cancel to close this message without transfering data.

OK   Cancel

RAND *TR433-D.23*

The user may also choose to enter the disutilities directly into the SLAM program on the Advanced Assignment Rule subform by clicking on the *Directly Edit* button. Once the disutilities have been entered in the Disutilities table, the user need only click on the *Done* button and the disutilities are saved for that assignment rule.

It is important to note that the example in this section and those that follow are for assignment rules that use only active and reserve forces with the maximum number of periods home/away set at 12 and 24, respectively. For scenarios using other forces and different maximum numbers of periods, a different spreadsheet will be created by the SLAM program, but the core principles remain the same (i.e., using average disutility per effective period and steady-state rotations to determine the disutilities). The following sections show how this base assignment rule can be adapted for different force allocation policies.

D.2. Using the Reserve Component More Frequently

The second analysis in Chapter Four varies reserve utilization policies. The reserve utilization policy used in the previous analyses limited deployments to 1 year in 6 (4 quarters in 24). The analyses in Section 4.5.2 look at three variations of this policy: 1 year in 5, 1 year in 4, and 1 year in 3. Each of these variants requires a new assignment rule. Since each variant requires similar changes, this section shows how to adapt the base assignment rule from Section D.1 (with reserve forces ineffective for the first two quarters of deployment and a 1-year-in-6 reserve rotation) to a reserve rotation of 1 year in 4, with reserves still ineffective for the first two quarters of deployment.

This assignment rule maintains the same disutilities for the first rotation of actives as the base assignment rule in Section D.1. However, the reserve disutilities need to be changed so that reserve units are deployed for 4 quarters out of every 16, rather than 4 out of every 24. Figure D.16 showed that, for the actives, the average disutility of a deployment of length 1 quarter for an active unit at home 8 quarters is 38 (the largest for the active rotation). Therefore, the average disutility of deployment for reserve units deployed 3 quarters from being at home 24 quarters must be greater than 38. This can be done by incrementing the largest ranking from the active rotation (38 for *Act/H-8*) by setting *Res/H-24* equal to 39. Because we want to deploy reserve units on a 4-year cycle (16 quarters), we increment the reserve time at home disutilities until we reach *Res/H-15* (in order from *Res/H-24*=39 to *Res/H-15*=48). After this, we continue to increment the disutilities, but we now get into a steady-state rotation, setting *Res/D-1*=49, *Res/H-14*=50, *Res/D-2*=51, and *Res/H-13*=52. These last disutilities are similar to what we did in Section D.1.5, except that here they are based on a 16-period maximum rotation rather than 24.

After incrementing to the point where forces are effective, the next step is to set *Res/D-3* so that the average disutility of deploying a reserve unit from at home 24 periods for 4 periods is larger than the average disutility of deploying a reserve unit from at home 13 periods for 3 periods. We found that setting *Res/D-3* equal to 200 accomplished this goal and followed this by incrementing *Res/H-12* to 201. The average disutility table for the reserves is shown in Figure D.24.

Figure D.24
Average Disutility per Effective Period for Reserve Deployments of 1 Year in 4

Res/H24	Res/H23	Res/H22	Res/H21	Res/H20	Res/H19	Res/H18	Res/H17	Res/H16	Res/H15	Res/H14	Res/H13	Res/H12
0.0	0.0	0.0	0.0	0.0	0.0	0.0	0.0	0.0	0.0	0.0	0.0	0.0
0.0	0.0	0.0	0.0	0.0	0.0	0.0	0.0	0.0	0.0	0.0	0.0	0.0
139.0	140.0	141.0	142.0	143.0	144.0	145.0	146.0	147.0	148.0	150.0	152.0	301.0
169.5	170.0	170.5	171.0	171.5	172.0	172.5	173.0	173.5	174.0	175.0	176.0	250.5

The last step in defining this assignment rule is to set the disutilities for *Act/H-7* through *Act/H-1*. We do this using exactly the same methodology as in the previous section. *Act/H-7* was set so that the average disutility of deploying an active unit at home 7 periods for 4 periods (513.3) was larger than that of deploying a reserve unit home 12 periods for 3 periods (301). The same process was used to bring actives from home with less than 7 periods at home, as shown in Figure D.25.

Figure D.25
Average Disutility for Reducing Active Time at Home Below 2 Years, with Reserves Used 1 Year in 4

Act/H7	Act/H6	Act/H5	Act/H4	Act/H3	Act/H2	Act/H1
2000.0	8000.0	32000.0	128000.0	512000.0	2048000.0	8192000.0
1002.0	4002.0	16002.0	64002.0	256002.0	1024002.0	4096002.0
672.0	2672.0	10672.0	42672.0	170672.0	682672.0	2730672.0
513.3	2013.3	8013.3	32013.3	128013.3	512013.3	2048013.3

RAND *TR433-D.25*

These are all the changes that need to be made to the base assignment rule to deploy reserves 1 year in every 4, maintain active and reserve deployments at 1 year each, and fill other requirements by reducing active time at home below 2 years. Similarly to the first section, we filled all unfilled disutilities with a large number (999,999,999) to make sure that those force-status-periods are never deployed. The Disutilities table for this assignment rule is shown in Figure D.26.

Figure D.26
Disutilities for Reserve 1-Year-in-4 Assignment Rule

Disutilities Table			
	Act/H	Act/D	Res/H	Res/D
1	8E+06	4	1E+09	41
2	2E+06	12	1E+09	51
3	512000	37	1E+09	200
4	128000	1E+09	1E+09	1E+09
5	32000	1E+09	1E+09	1E+09
6	8000	1E+09	1E+09	1E+09
7	2000	1E+09	1E+09	1E+09
8	38	1E+09	1E+09	1E+09
9	13	1E+09	1E+09	1E+09
10	5	1E+09	1E+09	1E+09
11	2	1E+09	1E+09	1E+09
12	1	1E+09	201	1E+09
13			52	1E+09
14			50	1E+09
15			48	1E+09
16			47	1E+09
17			46	1E+09
18			45	1E+09
19			44	1E+09
20			43	1E+09
21			42	1E+09
22			41	1E+09
23			40	1E+09
24			39	1E+09

RAND *TR433-D.26*

For the two assignment rules not described in this section (reserves used 1 year in 5 and 1 year in 3), the method is analogous to the one described. The only difference will be that the

1-year-in-5 assignment rule will require an initial increment from *Res/H-24* to *Res/H-19* and the 1-year-in-3 assignment rule will require an initial increment from *Res/H-24* to *Res/H-11*.

D.3. A Global War on Terror Assignment Rule

The base assignment rule discussed in Section D.1 is used for most of the analyses in Chapter Four of this document. In Chapter Five, a new assignment rule is developed to be more representative of the current deployment policies for the global war on terror. This assignment rule continues to limit active deployments to 4 quarters in every 12, but changes reserve deployments to 6 quarters in every 24 (from 4 in 24). All other requirements were met by reducing active time at home below 2 years. This section shows how to adapt the assignment rule in Section D.1 to take into account these longer reserve deployments.

Because of the similarities to the assignment rule defined in Section D.1, the initial disutilities for the actives remain identical. The only difference in this section is the disutilities of the reserves and the remaining active disutilities for dwell times under 2 years. Figure D.16 showed that, for the actives, the average disutility of a deployment of length 1 quarter for an active unit at home 8 quarters is 38 (the largest for the active rotation). Therefore, the average disutility of deployment for reserve units deployed 3 quarters from being at home 24 quarters must be greater than 38. This can be done by incrementing the largest ranking from the active rotation (38 for *Act/H-8*), setting *Res/H-24* equal to 39. We then set *Res/H-23* equal to 40. Following the same steady-state ordering used earlier, we next deploy a reserve unit for 2 periods by setting *Res/D-1* equal to 41. Next, we set *Res/H-22* equal to 42, *Res/D-2* equal to 43, and *Res/H-21* equal to 44.

After specifying these same initial incremental disutilities used in the previous assignment rule, we now focus on deploying a reserve unit for 6 periods (4 effective periods). In the steady state, a reserve unit deployed for 6 periods will be at home for 18 periods (for a 6-year cycle length).

We first want to set *Res/D-3* so that the average disutility of deploying a reserve unit for 4 periods from at home 24 periods is larger than that of deploying a reserve unit for 3 periods from at home 21 periods. We accomplished this by setting *Res/D-3* equal to 135. Next, we set *Res/H-21* by incrementing the disutilities to 136. The next comparison involves setting *Res/D-4* so that the average disutility of being deployed 5 periods from at home 24 periods is greater than that of being deployed 3 periods from at home 20 periods. We accomplished this by setting *Res/D-4* equal to 405, and then set *Res/H-19* to 406 by incrementing the disutilities. We next set *Res/D-5* so that the average disutility of being deployed 6 periods from at home 24 is greater than that of being deployed 3 periods from at home 19 periods. To do this, we set *Res/D-5* equal to 1,305, and therefore also set *Res/D-18* equal to 1,306. Figure D.27 shows part of the Average Disutility per Effective Period table for the reserves used to make all of the calculations discussed in this section.

Figure D.27
Average Disutility per Effective Period for Reserves on Longer Deployments

Res/H24	Res/H23	Res/H22	Res/H21	Res/H20	Res/H19	Res/H18
0.0	0.0	0.0	0.0	0.0	0.0	0.0
0.0	0.0	0.0	0.0	0.0	0.0	0.0
123.0	124.0	126.0	128.0	220.0	490.0	1390.0
129.0	129.5	130.5	131.5	177.5	312.5	762.5
221.0	221.3	222.0	222.7	253.3	343.3	643.3
492.0	492.3	492.8	493.3	516.3	583.8	808.8

RAND *TR433-D.27*

The final step in specifying this assignment rule is to specify *Act/H-7* through *Act/H-1*. We want to set these so that once the actives are deployed 4 quarters in 12 and reserves are deployed 6 in 24, the assignment rule will deploy active units at home for shorter periods of time to meet any further requirements. We begin by setting *Act/H-7* so that the average disutility of deploying an active unit 4 periods from at home 7 periods is greater than the largest average disutility from the reserve deployment box (in this case, *Res/H-18:D-3*, with an average disutility of 1,390). We set *Act/H-7* equal to 6,000 so that the average disutility of *Act/H-7:D-4* equals 1,513. Next, we want to ensure that the average disutility of deploying an active unit for 4 periods from at home 6 periods is larger than that of deploying an active unit at home 7 periods for 1 period. We did this by setting *Act/H-6* equal to 24,000. We repeated the same process to set *Act/H-5* through *Act/H-1* to create the Average Disutility per Effective Period table shown in Figure D.28.

Figure D.28
Using Actives on Short Reset Times for GWOT Assignment Rule

Act/H7	Act/H6	Act/H5	Act/H4	Act/H3	Act/H2	Act/H1
6000.0	24000.0	96000.0	384000.0	1536000.0	6144000.0	24576000.0
3002.0	12002.0	48002.0	192002.0	768002.0	3072002.0	12288002.0
2005.3	8005.3	32005.3	128005.3	512005.3	2048005.3	8192005.3
1513.3	6013.3	24013.3	96013.3	384013.3	1536013.3	6144013.3

RAND *TR433-D.28*

Lastly, we assigned a very large disutility (999,999,999) to all force periods that we have not yet given disutilities to, to make sure that these force-status-periods are never deployed. The Disutilities table for this assignment rule is shown in Figure D.29.

Figure D.29
Disutilities Table for GWOT Assignment Rule

Disutilities Table				
	Act/H	Act/D	Res/H	Res/D
1	2E+07	4	1E+09	41
2	6E+06	12	1E+09	43
3	2E+06	37	1E+09	135
4	384000	1E+09	1E+09	405
5	96000	1E+09	1E+09	1305
6	24000	1E+09	1E+09	1E+09
7	6000	1E+09	1E+09	1E+09
8	38	1E+09	1E+09	1E+09
9	13	1E+09	1E+09	1E+09
10	5	1E+09	1E+09	1E+09
11	2	1E+09	1E+09	1E+09
12	1	1E+09	1E+09	1E+09
13			1E+09	1E+09
14			1E+09	1E+09
15			1E+09	1E+09
16			1E+09	1E+09
17			1E+09	1E+09
18			1306	1E+09
19			406	1E+09
20			136	1E+09
21			44	1E+09
22			42	1E+09
23			40	1E+09
24			39	1E+09

RAND *TR433-D.29*

We have now fully defined this assignment rule up to the point that we need to use it for the analyses in this document. Again, it is up to the user how far he or she wishes to define the assignment rule. To save time and effort, we stopped here since we knew that this would be sufficient.

D.4. Other Assignment Rules

The assignment rules defined in this appendix show the methodology for developing an assignment rule with forces that are not fully effective in all periods. There are many varieties of assignment rule, and not all of them can be covered in this document. The important lesson from these assignment rules is that the user should be concerned with average disutility per effective period (not directly with disutility) and steady-state deployment patterns. The spreadsheets for these examples are included with the SLAM distribution package. These spreadsheets can be opened directly in Excel or from the SLAM program when the user has open one of the back-end databases included in the distribution package. An assignment rule spreadsheet will remain associated with the assignment rules in the back-end database as long as there are no changes made to the name of the back-end database or the name of the assignment rule (either in the back-end or in the file name of the spreadsheet).

The assignment rules examined here share many of the same parameters. For assignment rules using different number of periods home versus away or different numbers or types of forces, the spreadsheets will look different, but the methodology remains the same.

D.5. Validating Assignment Rules

Before using an assignment rule to perform a force structure analysis, the user may wish to test the assignment rule to confirm that it is deploying forces as desired. This can be done by setting up test runs and using the Force Assignment report to examine the results. The user can set up a test run by specifying a contingency with a constant requirement and then using the Force Assignment report to see how forces are deployed to meet that requirement. The user can then create other test runs with requirement levels greater and less than the level of the first test run to examine the order in which forces are being deployed.

As an example, to test the *BaseActiveReserve* assignment rule specified in this appendix, we could create a test run that includes a contingency with a requirement of 16 units and a force size of 48 active units and 48 reserve units. When we examine the Force Assignment report for this run, we would expect only active forces to be deployed for 4-period deployments, followed by 8-period dwell times. We could then create a run with the same force size but a requirement of 20 units. For this run, we would expect active deployments to remain the same as the previous run but that reserve units would be deployed for 4 periods (2 periods effective) with 20-period dwell times. Lastly, we may wish to create one more run with the same force size but with a lower requirement of 12 units. For this run, we expect no reserves to be deployed and active units to be deployed for 3 periods, with 9-period dwell times.

The user can create any number of test runs to ensure that the assignment rule is working properly. However, for the analyses in this report, we have found that if one follows the assignment rule methodology described in this appendix, that force assignment is always done as expected. But, if the user begins to create assignment rules with a greater number of forces (more than just actives/reserves), a different number of ineffective periods for reserves (or other forces), or different maximum home/deployed times, we suggest that the user test out these assignment rules first before using them in an analysis.

Tables and Modules

This appendix provides a listing of all of the tables and modules in the SLAM program, followed by brief descriptions.

E.1. Tables

The backbone of the SLAM program is the set of tables that contain all of the data. We suggest that the analyst not directly access a table unless absolutely necessary, as this may lead to corruption of the data. Experienced Microsoft Access users may find some benefit in looking at and manipulating the tables. Table E.1 provides a brief description of each table in the SLAM program.

Table E.1
SLAM Tables

Table Name	Description
tblAdmin	This table retains the name of two databases: (1) the default database name, and (2) the name of the last database used. This information is used to populate the SLAM start-up screen. This table also includes the location that the user selects for the simulation to run (local or remote machine), the working directory for GAMS to run in, and the per-period run-time limit for GAMS (all of these can be changed by the user from the SLAM Options form).
tblAPMtemp	This is a temporary table that is used when the analyst changes the order in which units are assigned. A temporary table is needed to preserve the previous ordering while updating the tblAssignmentRulePeriodMatrix table with the new ordering.
tblAssignmentRule	This table contains general information for each assignment rule.
tblAssignmentRulePeriodMatrix	This table contains the user-defined disutilities that determine the order in which units are assigned.
tblBasicAssignmentRules	This table includes a simplified version of the assignment rule period matrix, in which force types can be grouped for ease of use (only useful when assigning forces that are fully effective in all periods).
tblContingency	This table contains general information for each contingency.
tblContingencyStates	This table lists all of the states for each possible contingency. It also lists the total number units required (for each of the six requirements) for each state of each contingency.

Table E.1—Cont'd

Table Name	Description
tblContingencyTransMatrix	This table contains the transition probabilities from each state to every other state (e.g., *War* to *Peace*, *Peace* to *Stabilization*) for each contingency.
tblForcePeriodEffective	This table contains the various types of effectiveness for each force over all periods.
tblForcePeriodTraining	This table contains the training value of each type of training for each force over all periods.
tblForces	This table contains the general characteristics of each force type, including costs and maximum number of periods home and deployed.
tblRun	This table contains all of the run parameters for each run. It includes fields for the number of periods, the assignment rule, the various look-ahead parameters, training capacities, and the solution methodology (linear or integer program).
tblRunContingencies	This table includes an entry for each contingency included in each run along with its initial state.
tblRunContingencyPeriod	*Output*: This table provides the state of every contingency in every period for each simulation run. This table is populated by the output from the Markov process.
tblRunForceLevels	This table identifies each force type used for each run. It also defines the number of available forces for each force type in each run.
tblRunForcePeriod	*Output*: This table includes the number of units in each state in every period for each simulation run. This table is populated by the output from the linear program (the stock variable; see Appendix C).
tblRunForceTransitionStress	This table allows the user to associate stress measures with specific transitions made in tblRunSolutions.
tblRunMeasures	*Output*: This table records the start time, end time, and total run time of each run, in minutes. It also keeps track of the number of periods, solution type (integer or linear), and number of look-ahead periods.
tblRunSolutions	*Output*: This table includes the linear program solutions for each time period (the X variable; see Appendix C).
tblRunStress	This table associates each of the four stress measures with a force status. Stress measures are initially unassociated.
tblStateList	A table listing each state and state abbreviation. This table can be used to add or remove states.
tblStressMeasures	This table lists all of the stress measures for each simulation run. In this version of the SLAM model, these are static; there are only four stress measures (this table is not used in this version of the SLAM program but may be utilized in future versions).

E.2. Modules

The SLAM program consists of seven modules: Execute, GAMSCode, GAMSCodeLP2, GAMSCodeLP2ToGAMS, OpenDialogBox, ShellWaitCode, and SLAMFunctions. The modules and their associated functions are described below.

- **Execute**: This contains the code that runs the SLAM program by simulating contingencies and solving the linear program. This version of the SLAM program uses Microsoft Excel and Frontline Solver, rather than GAMS (not used in current version of SLAM program), to solve the linear programs.
- **GAMSCode**: This module calls GAMS iteratively to solve a simple linear program in each period. This module is not used in this model because it solves only very simple linear programs.
- **GAMSCodeLP2**: This module calls GAMS iteratively to solve the linear programs that assign forces in the SLAM program. This module is no longer used because it calls GAMS during each period, which increases run times substantially.
- **GAMSCodeLP2ToGAMS**: This module calls GAMS once at the beginning of the run and reads in the GAMS output files at the end. This module is used for all of the analyses contained in this report.
- **OpenDialogBox**: This contains code that opens the Windows Open File dialog box.
- **ShellWaitCode**: This module contains code that allows Microsoft Access to wait for GAMS to finish before moving to the next task.
- **SLAMFunctions**: This contains seven separate public functions that can be called from anywhere:
  - *Checkperiods*
  - *FindMax*
  - *Openfile*
  - *LinkEm*
  - *MarkEmNoEdit*
  - *Sleep*
  - *Savefile.*

Bibliography

Center for Army Analysis, *Stochastic Analysis of Resources for Deployments and Excursions*, Fort Belvoir, Va., July 1999.

———, *The Army Rotation Rule Project*, Fort Belvoir, Va., June 2001.

———, *Deployment Tempo Analysis in the U.S. Army*, Fort Belvoir, Va., October 2002.

———, *Alternative Deployment Duration—Reserve Component*, Fort Belvoir, Va., October 2003.

Congressional Budget Office, *An Analysis of the U.S. Military's Ability to Sustain an Occupation of Iraq*, Washington, D.C., September 2003.

———, *Options for Restructuring the Army*, Washington, D.C., May 2005.

Davis, Lynn E., J. Michael Polich, William M. Hix, Michael D. Greenberg, Stephen D. Brady, and Ronald E. Sortor, *Stretched Thin: Army Forces for Sustained Operations*, Santa Monica, Calif.: RAND Corporation, MG-362-A, 2005. As of July 31, 2007:
http://www.rand.org/pubs/monographs/MG362/

Hillier, Frederick S., and Gerald J. Lieberman, *Introduction to Operations Research*, 8th ed., Boston: McGraw-Hill, 2005.

Krepinevich, Andrew, *The Thin Green Line*, Washington D.C.: Center for Strategic and Budgetary Assessments, 2006.

Larson, Eric V., David T. Orletsky, and Kristin Leuschner, *Defense Planning in a Decade of Change: Lessons from the Base Force, Bottom-Up Review, and Quadrennial Defense Review*, Santa Monica, Calif.: RAND Corporation, MR-1387-AF, 2001. As of August 18, 2007:
http://www.rand.org/pubs/monograph_reports/MR1387/

O'Hanlon, Michael, *Defense Strategy for the Post-Saddam Era*, Washington D.C.: Brookings Institution Press, 2005.